PSpice for Filters and Transmission Lines

Paul Tobin
School of Electronic and Communications Engineering
Dublin Institute of Technology
Ireland

SYNTHESIS LECTURES ON DIGITAL CIRCUITS AND SYSTEMS #8

ABSTRACT

In this book, *PSpice for Filters and Transmission Lines*, we examine a range of active and passive filters where each design is simulated using the latest Cadence Orcad V10.5 PSpice capture software. These filters cannot match the very high order digital signal processing (DSP) filters considered in *PSpice for Digital Signal Processing*, but nevertheless these filters have many uses. The active filters considered were designed using Butterworth and Chebychev approximation loss functions rather than using the 'cookbook approach' so that the final design will meet a given specification in an exacting manner. Switched-capacitor filter circuits are examined and here we see how useful PSpice/Probe is in demonstrating how these filters, filter, as it were. Two-port networks are discussed as an introduction to transmission lines and, using a series of problems, we demonstrate quarter-wave and single-stub matching. The concept of time domain reflectrometry as a fault location tool on transmission lines is then examined. In the last chapter we discuss the technique of importing and exporting speech signals into a PSpice schematic using a tailored-made program Wav2ascii. This is a novel technique that greatly extends the simulation boundaries of PSpice. Various digital circuits are also examined at the end of this chapter to demonstrate the use of the bus structure and other

KEYWORDS

Passive filters, Bode plotting, Active filter design, Multiple feedback active filter, Biquad active filter, Monte Carlo analysis, Two-Port networks, z-parameters.

I dedicate this book to my wife and friend, Marie and sons Lee, Roy, Scott and Keith and my parents (Eddie and Roseanne), sisters, Sylvia, Madeleine, Jean, and brother, Ted.

Contents

Preface

In book 1, PSpice for Circuit Theory and Electronic Devices, we explain in detail the operational procedures for the new version of PSpice (10.5) but I include here a very quick explanation of the project management procedure that must be followed in order to carry out even a simple simulation task. Before each simulation session, it is necessary to create a project file using the procedure shown in Figure 1. This will not be mentioned each time we consider a new schematic as it becomes tedious for the reader seeing the same statement "Create a project called Figure 1-008.opj etc" before each experiment. After selecting Capture CIS from the Windows start menu, select the small folded white sheet icon at the top left hand corner of the display as shown.

Enter a suitable name in the **Name** box and select **Analog or Mixed A/D** and specify a **Location** for the file. Press **OK** and a further menu will appear so tick **Create a blank project** as shown in Figure 2.

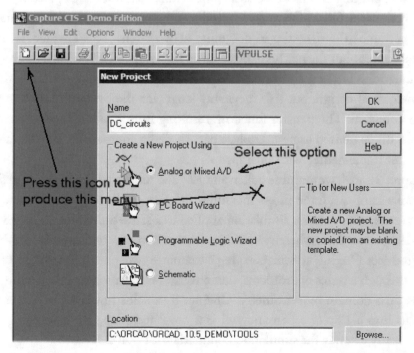

FIGURE 1: Creating new project file

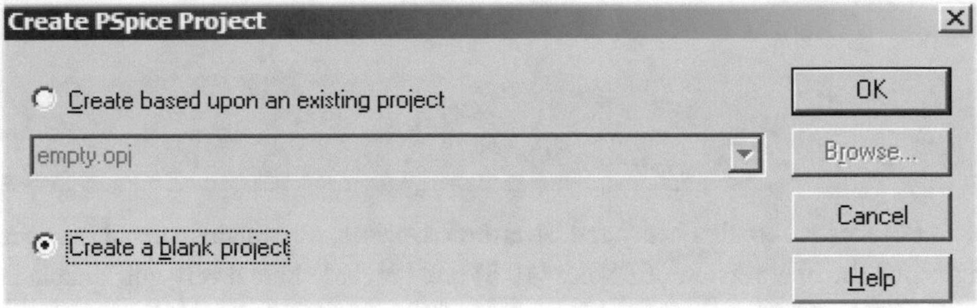

FIGURE 2: Create a blank project

In the project management area, expand the **DC_circuits** directory (or whatever you called the project) to produce an empty schematic area called **Page 1** where components are placed. Libraries have to be added, (**Add library**) by selecting the little **AND** symbol in the right toolbar icons. The easiest method is to select all the libraries. However, if you select **Create based upon an existing project**, then all previously used libraries associated with that project will be loaded.

Chapter 1 looks at passive filters and introduces the concept of roll-off rate and cut-off frequency. Bode plotting is considered as an approximate technique for plotting the frequency response of first-order filters. Chapter 2 introduces approximation loss functions to produce a transfer function as the first step in designing active filters. This chapter has quite a high mathematical content but the mathematics may be side-stepped if you wish to proceed directly with the simulation. There is always the temptation for students to simulate without doing a proper mathematical design, but this inevitably lengthens the overall simulation process and should be discouraged. The transfer functions resulting from this loss function may have their amplitude responses plotted using a Laplace part so that a comparison may be made with the final active filter implemented with opamps.

In chapter 3 and 4, we examine Sallen and Key and multiple feed-back active filters, whilst chapter 5 looks at bi-quadratic filters including the state-variable type and we demonstrates Monte Carlo sensitivity analysis/simulation applied to a sixth-order filter. Chapter 6 discusses the switched-capacitor filter whereby switching charge between two capacitors mimics a resistance and produces filtering action. Chapter 7 examines two-port networks and transmission lines where, through a series of problems, demonstrates transmission line matching techniques and explains time domain reflectrometry. Chapter 8 explains how we can import and export speech signals into a PSpice schematic and is a novel technique for examining realistic test signals that greatly extends the simulation boundaries of PSpice.

ACKNOWLEDGMENTS

Anyone who has written a textbook knows that a successful product relies on lots of people not directly involved. I would like to thank a retired head of our department, Bart O'Connor whose lectures I had the pleasure of attending all those years ago on transmission line theory. Thanks to my son, Lee Tobin, who wrote a very useful program 'Wav2ASCII' for creating ASCII speech files for processing in PSpice and subsequently playing back the processed file after simulation. This small program really extends the simulation boundaries of PSpice greatly. Lastly, thanks to Dr Mike Murphy, Director of Engineering, Dublin Institute of Technology (DIT) for his interest and encouragement in my PSpice books.

CHAPTER 1

Passive Filters and Bode Plotting

1.1 FILTER TYPES

Filters are used to modify the frequency spectrum of signals and the five basic filter types are:

- Low-pass . . . . passes low frequencies but attenuates high frequencies,

- High-pass . . . . passes high frequencies but attenuates low frequencies

- Bandpass. . . . . passes a band of frequencies and attenuates frequencies outside that band

- Bandstop. . . . . attenuates a band of frequencies but passes all other frequencies outside this band

- All-pass . . . . not really a filter but used to modify the phase response

An ideal "brick-wall" low-pass filter response is shown superimposed on a "real" low-pass filter response in Fig. 1.1. What this brick-wall filter does is to pass all frequencies up to a certain frequency called the *cut-off frequency* ω_p, where the subscript stands for passband (also called $\omega_c = 2\pi f_c$, the subscript stands for cut-off frequency), and then infinitely attenuate all frequencies past this value. A real low-pass filter response, however, passes all frequencies with very little attenuation from DC up to ω_p but gradually attenuates signals above this frequency. In a simple *CR* low-pass filter response, the signal is gradually attenuated from DC but, because of the way capacitive reactance changes with frequency, the attenuation does not become apparent until we approach the cut-off frequency. At low frequencies the output signal is in step with the input signal, i.e., no phase difference between them, but at high frequencies, the output lags behind the input by ninety degrees, i.e., $-90°$, i.e., there is a delay between them.

1.2 DECIBELS

The decibel is the normal unit of measurement in filter design. The logarithmic nature of the human ear response thus makes the dB a natural choice for tone control filters such as the bass/treble filters in audio amplifiers. We must consider the Bel as the fundamental unit and is

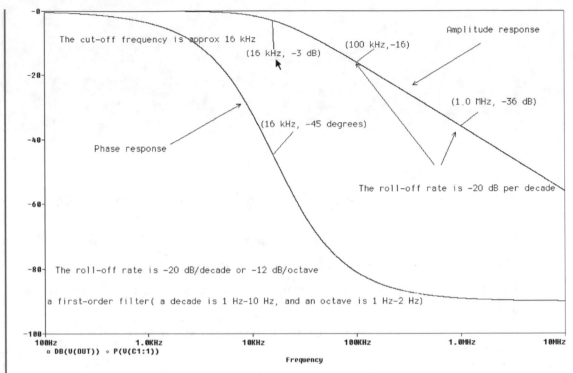

FIGURE 1.1: Brick-well filter response and actual response

a logarithmic ratio of input and output powers defined as

$$\text{Bel} = \log_{10} \frac{P_{\text{out}}}{P_{\text{in}}}. \tag{1.1}$$

However, the Bel is too large for most applications, so the decibel is used instead. (Imagine the inconvenience of expressing your height in fractions of a kilometer.) There are ten decibels in a Bel so the dB is

$$\text{dB} = 10 \log_{10} \frac{P_{\text{out}}}{P_{\text{in}}}. \tag{1.2}$$

1.2.1   The Cut-Off Frequency

To demonstrate decibels and the important concept of cut-off frequency, consider an audio amplifier producing 10 Watts of output power, P_{out} measured in the midband audio region. This amplifier produces less output power at very low and very high frequencies. At a certain frequency the output power gradually reduces and at a certain frequency, f_p, called the upper cut-off frequency, the output power is measured as 5 W. This power ratio of 0.5 is called the half-power point and defines the cut-off frequency (also known as the corner/break/transition frequency). The ratio of the power at f_p to the power delivered over the passband frequency range can be expressed as

$$\text{dB} = 10 \log_{10} \frac{P_{\text{out}}}{P_{\text{in}}} = 10 \log_{10} \frac{5}{10} = 10 \log_{10} 0.5 = -3.01\,\text{dB} \approx -3\,\text{dB}. \qquad (1.3)$$

This particular attenuation is called the passband edge frequency attenuation A_{max} (explained later). We may express each power component in (1.3) as V^2/R, thus the ratio in dB is

$$\text{dB} = 10 \log_{10} \frac{V_{\text{out}}^2 / R_{\text{out}}}{V_{\text{in}}^2 / R_{\text{in}}}. \qquad (1.4)$$

Maximum power transfer between a source and a load is desirable in electronic systems [1]. Examples are: a microphone 50 kΩ source impedance connected to an amplifier input impedance of 50 kΩ, a TV whose input impedance of 75 Ω is connected to a transmission line whose characteristic impedance is 75 Ω, phone handset impedance (300 Ω) to the transmission line characteristic impedance (300 Ω), etc. In these examples, we see that when the source resistance is equal to a load resistance, (1.4) is simplified:

$$\text{dB} = 10 \log_{10} \left(\frac{V_{\text{out}}}{V_{\text{in}}} \right)^2 = 20 \log \frac{V_{\text{out}}}{V_{\text{in}}}. \qquad (1.5)$$

A power halving is -3 dB but we wish to see what this is in terms of a voltage ratio. This is calculated by taking the antilog of (1.5) as a $\log(-3/20) = 0.707 = 1/\sqrt{2}$. The passband region, DC to f_p, in a low-pass filter amplitude response, is constant, with little or no attenuation up to the cut-off frequency. When the resistance is equal to the reactance then this frequency makes the transfer function $V_{\text{out}}/V_{\text{in}}$ equal to TF $= 1/\sqrt{2} = 0.707$. However, the passband gain may not always be 1, so care has to be exercised. For example, the passband gain could be 4, hence the transfer function would be TF $= 4/\sqrt{2}$. Note, however, a voltage doubling in dB is $20 \log_{10} 2 = 6$ dB.

1.2.2 Filter Specification

A low-pass filter specification includes the passband edge frequency attenuation A_{max} in dB. This is the attenuation a signal will experience at the passband edge frequency f_p, in a simple CR filter. The attenuation at the stopband edge frequency, f_s, is A_{min}, which, for the first-order CR filter, is -20 dB. These parameters are all shown on the low-pass filter response shown in Fig. 1.2. The parameters are the same for a high-pass filter but a mirror image of the low-pass filter. The *transition region* between the passband and stopband regions reduces in width when the filter order is increased. For example, a first-order filter is achieved using a capacitor and

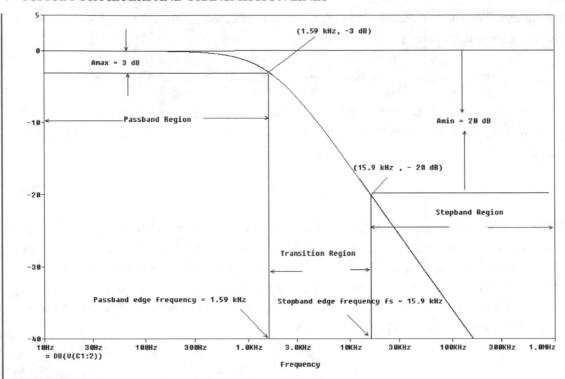

FIGURE 1.2: Low-pass filter response and specification parameters

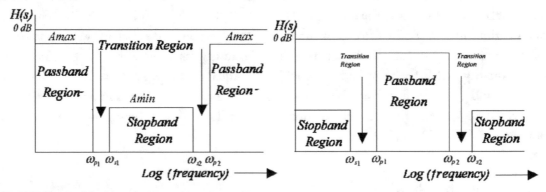

FIGURE 1.3: Ideal bandstop and bandpass filter responses

resistor connected as a two-port network. In a high-pass filter, all frequencies below the cut-off frequency are attenuated at −20 dB/decade, but frequencies above the cut-off frequency are passed with little attenuation.

Ideal bandpass and band-stop filter responses are shown in Fig. 1.3, where the bandpass filter passes all frequencies over a desired band but attenuates all other frequencies outside this

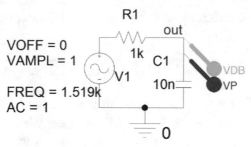

FIGURE 1.4: LPF and probe-marker correlation

band. In a bandstop filter, signals are attenuated over a band of frequencies but outside this band are passed with little attenuation.

The specification for bandpass and bandstop filters must include two frequency pairs.

1.3   TRANSFER FUNCTION: FIRST-ORDER LOW-PASS *CR* FILTER

We may examine the behavior of the filter shown in Fig. 1.4 by considering two extreme frequencies. At 0 Hz (DC), the capacitor reactance has an infinite value (an open circuit), hence the output voltage is the same as the input voltage, i.e., no attenuation because there is no current flowing in the circuit so no voltage drop across the resistance. At infinity frequency, the capacitor reactance is zero (a short circuit) hence the output voltage is zero. So we can surmise that the device is a low-pass filter. However, this is only a rough picture of what goes on and we need to fill the gaps frequency-wise as it were.

A filter transfer function (TF) relates the output and input voltages and we need to get an expression that relates the output voltage across the capacitor to the input voltage. We could write the TF directly using the potential divider method but we will derive it from the basic circuit theory. The output voltage by Ohm's law is

$$V_2 = \text{current} \times \text{reactance} = i(-jX_c) = \frac{i}{j\omega C}. \tag{1.6}$$

Substituting the circuit current $i = \frac{\text{total voltage}}{\text{total impedance}} = \frac{V_1}{R + 1/j\omega C}$ into (1.6) yields

$$V_2 = \frac{V_1}{R + 1/j\omega C}\frac{1}{j\omega C} \Rightarrow \frac{V_2}{V_1} = \frac{1}{1 + j\omega CR}. \tag{1.7}$$

A good rule of thumb when deriving transfer functions is to eliminate any fractional quantities by multiplying by the denominator of the fraction such as $j\omega C$ in (1.8). We could, of course, have written this transfer function by applying the voltage divider rule as:

$$H(s) = \frac{V_2}{V_1} = \frac{1/j\omega C}{R + 1/j\omega C} \times \frac{j\omega C}{j\omega C} = \frac{1}{1 + j\omega CR} = \frac{1}{1 + j2\pi fCR} = \frac{1}{1 + j\omega/\omega_p}. \quad (1.8)$$

Here $1/CR$ is the cut-off frequency ω_p. So when $\omega = \omega_p$ the $TF = 1/(1 + j1) \Rightarrow |TF| = 1/\sqrt{2} = 0.707$. A complex number, $Z = R + jX$, has a magnitude and phase defined as $Z = \sqrt{R^2 + X^2} \angle \tan^{-1}(X/R)$. Thus, the LPF transfer function $1/(1 + j\omega CR)$ has a magnitude $1/\sqrt{1^2 + \omega^2 C^2 R^2}$, unity gain at DC (the passband region) and the phase response plotted using $\theta = -\tan^{-1} \omega CR$. We already introduced the important concept of cut-off frequency when we considered the power relationships in the previous section but another visit should reinforce this important concept. This special frequency occurs where the real part of the transfer function equals the imaginary part. To obtain an expression for ω_p, we equate the TF magnitude to $1/\sqrt{2} = 0.707$. Thus, we may write

$$|TF| = \frac{V_o}{V_{\text{in}}} = \frac{1}{\sqrt{1 + \omega^2 C^2 R^2}} = \frac{\text{passband gain}}{\sqrt{2}} = \frac{1}{\sqrt{2}} \Rightarrow \sqrt{1 + \omega^2 C^2 R^2} = \sqrt{2}. \quad (1.9)$$

Squaring both sides $\omega^2 C^2 R^2 + 1 = 2 \Rightarrow \omega^2 C^2 R^2 = 1$ yields the cut-off frequency, in radians per second, as

$$\omega = 1/CR = \omega_p. \quad (1.10)$$

The subscript tells us that at this frequency, the output signal power is reduced by 3 dB (or the voltage reduced to 0.707 V_{in}). The cut-off frequency in Hz is $\omega_p/2\pi$, so the cut-off frequency for $C = 10$ nF and $R = 10$ kΩ is

$$f_p = \frac{1}{2\pi CR} = \frac{1}{2\pi \left(10 \times 10^{-9}\right)\left(10 \times 10^3\right)} = 1.59 \text{ kHz}. \quad (1.11)$$

1.3.1   Bode Plotting

Bode plotting (Hendrik Bode—1905–1982) is a useful graphical method for plotting the approximate filter amplitude and phase responses. We do this by splitting the magnitude of the TF into smaller parts containing terms of the form $1 \pm j\omega/\omega_p$ and plotting each TF part using straight-line segments. The complete frequency response is then obtained by adding the individual responses together. The TF is manipulated into a form $(1 \pm j\omega/\omega_p)$ by making the real part of the complex number one. This is done by dividing every term in the transfer function by the real part of the complex number. We then get the magnitude of each part of the TF and express it in dB so that we can perform Bode analysis using straight-line or asymptotic

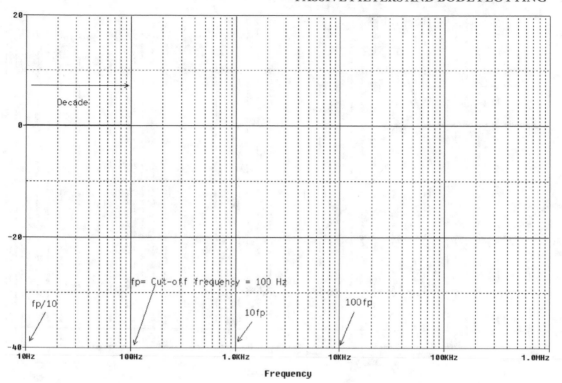

FIGURE 1.5: Part (a)

line segments. The transfer function in polar form is

$$TF = \frac{1}{1 + j\omega/\omega_p} = \frac{1}{1 + jf/f_p} \Rightarrow |TF| = \frac{1}{[1 + (f/f_p)^2]^{1/2}} \angle - \tan^{-1}(f/f_p). \quad (1.12)$$

If the cut-off frequency is 100 Hz, then the response is plotted two decades above and below this value, i.e., 1 Hz to 10 kHz. The numerator and denominator of the TF in (1.12), in dB:

$$|TF|_{dB} = 20\log(1) - 20\log\left[1 + \left(\frac{f}{f_p}\right)^2\right]^{1/2} = 0 - 10\log\left[1 + \left(\frac{f}{f_p}\right)^2\right]. \quad (1.13)$$

Evaluate the transfer function a decade above and below the cut-off frequency, i.e., at $f = f_p/10$, $f = 10f_p$.

a) The TF a decade below the cut-off frequency $f = f_p/10$ is

$$TF_{dB} = -10\log\left[1 + \left(\frac{f_p/10}{f_p}\right)^2\right] = -10\log[1 + (1/100)] \approx 0 \text{ dB}. \quad (1.14)$$

Plot this on log/linear paper as shown in Fig. 1.5.

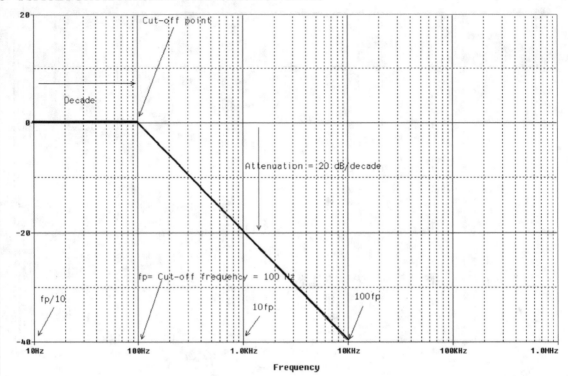

FIGURE 1.6: Part (a) + part (b)

b) For a decade above the cut-off frequency, $f = 10 f_p$, the transfer function in dB is

$$TF_{dB} = -10 \log \left[1 + \left(\frac{10 f_p}{f_p} \right)^2 \right] = -10 \log[1 + 100] \approx -20 \text{ dB/decade.} \quad (1.15)$$

Thus, for every decade increase in frequency, the output voltage falls by 20 dB. This is stated as −20 dB per decade increase in frequency, where a decade is from 1 kHz to 10 kHz, or 10 kHz to 100 kHz. This is equivalent to −6 dB per octave increase in frequency, an octave being a doubling in frequency, e.g., 1 kHz to 2 kHz. Fig. 1.6 shows a complete transfer function response. The y-axis is in dB and the x-axis is logarithmic to encompass a large frequency range. The transfer function is not evaluated at the cut-off frequency in a Bode plot, but may be calculated, however, to reinforce your understanding of the cut-off frequency concept.

c) Substitute f_p into the general frequency variable f to give

$$|TF|_{dB} = -10 \log \left[1 + \left(\frac{f_p}{f_p} \right)^2 \right] = -10 \log[1 + 1] = -3 \text{ dB.} \quad (1.16)$$

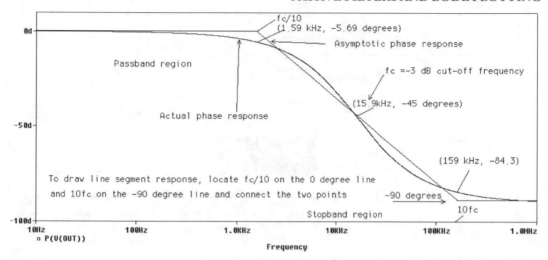

FIGURE 1.7: Asymptotic and actual phase response

The complete asymptotic response is obtained by adding the straight-line (Bode) responses together.

The filter order (the number of poles—more about this later) determines the roll-off rate in the transition region, i.e., a first-order has a roll-off rate of -20 dB/decade, a second-order is -40 dB/decade, etc. Active filters considered in later chapters have orders that rarely exceed 10, but digital filters can have very high orders [ref: 2].

1.3.2   Low-Pass CR Filter Phase Response

We need to plot the TF phase response as well as the magnitude part using the same Bode technique. The phase part of the transfer function is

$$TF(\theta) = -\tan^{-1}(f/f_p). \tag{1.17}$$

We may plot the approximate phase response by considering the value of the phase a decade above and below the cut-off frequency as before, i.e., $f = f_p/10$, and $f = 10f_p$. For example,

- At $f = f_p/10$, the phase is $\phi = -\tan^{-1}(f_p/10f_p) = -\tan^{-1}0.1 \approx 0°$
- At $f = f_p$, the phase $\phi = -\tan^{-1} f_p/f_p = -\tan^{-1}1 = -45°$
- At $f = 10f_p$, the phase is $\phi = -\tan^{-1}(10f_p/f_p) = -\tan^{-1}10 = -90°$.

The actual and asymptotic phase responses are shown in Fig. 1.7.

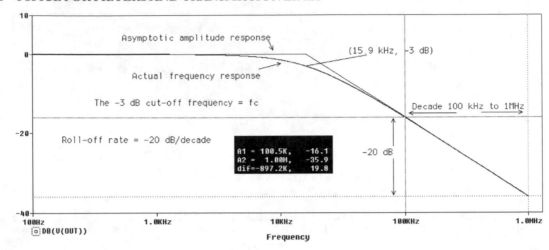

FIGURE 1.8: Low-pass filter response

FIGURE 1.9: Probe icons

1.4 MARKERS

Place decibel and phase markers from the **PSpice/Markers/Advanced** menu. Each marker will have a different color corresponding to the Probe output plot after simulation, which makes it handy for signal-marker identification. In this example, the output signal marker is red so the trace will be red too. Other available markers are group delay and real and imaginary parts of current/voltage. An alternative to using the phase marker is to enter **ATAN(IMG(V2(C1))/R(V2(C1)))*180/3.14159** in the **Trace Expression** box (i.e., $\tan^{-1}$(imag/real)*180/pi). The 180/pi is necessary to plot in degrees and not radians. Set the Analysis tab to Analysis type: AC Sweep/Noise, **AC Sweep Type to Logarithmic, Start Frequency = 100, End Frequency = 1000k, Point/Decade = 1000.** Select the blue right triangle, or press **F11** to produce the filter response shown in Fig. 1.8 where the horizontal frequency scale is plotted on a log basis.

For all Probes outputs, important information should be added, e.g., the cut-off frequency and roll-off rate. Fig. 1.9 shows the cursor icons bar (max value, min value, etc.). Locating the mouse pointer over each icon produces a small information textbox telling you the function of each one. From left to right from the Cursor Peak arrow: the first icon positions the cursor to the next local maximum or peak. The second icon positions the cursor to the next local minimum, and the third icon is the cursor slope. The fourth is the absolute minimum icon and

FIGURE 1.10: "search for min"

the next icon measures the cursor maximum. Pressing ⬚ will place x and y axis values at the cursor location.

Other icons are: Log x-axis, Fast Fourier Transform **FFT**, Performance Analysis, Log y-axis, Add Trace, Evaluate Function, and Add Text. The search command (shortcut **alt TCS**) is used to position the cursor at a specific place along a trace. The syntax is: Search [direction] [/start_point/] [#consecutive_points#] [(range_x[,range_y])]. Use the F1 button for further information on this topic. Turn on the cursor icon and select **Trace/Cursor/Search commands menu** to bring up Fig. 1.10 which shows the search command "**search for min**" entered in order to move the cursor to the minimum point on the response curve.

The icons at the start are the icons for examining in detail sections of the plot. Press the third icon from the left and then using the left mouse button expand a section of the plot.

1.4.1   Cut-Off Frequency and Roll-Off Rate

Two important things to measure from the frequency response are: (1) The **roll-off rate**—this is measured using two cursors separated a decade apart on the linear portion of the response (e.g., 10 kHz to 100 kHz but not on the curved portion of the response). Alternatively, you may measure it over an octave (e.g., 2 kHz to 4 kHz). (2) The **cut-off frequency**—use a single cursor to read the frequency where the output is down by 3 dB from the maximum value (midband or passband region). The passband gain is generally 0 dB for passive networks, but it may be nonzero for active networks which have gain. In that case read the frequency at the point where the level is 3 dB below the maximum passband gain.

1.5   ASYMPTOTIC FILTER RESPONSE PLOTTING USING AN FTABLE PART

Fig. 1.11 shows an **FTABLE** part used to produce a low-pass filter response in Bode asymptotic form, where the cut-off frequency is 1.59 kHz.

Select the **FTABLE** part and enter the table of values into the columns as shown in Fig. 1.12.

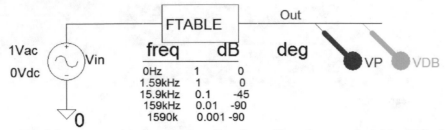

To plot response in magnitude rather than dB, enter mag in table attributes

FIGURE 1.11: Using the Ftable part to plot a Bode response

Reference	FTABLE2
ROW1	0Hz 1 0
ROW2	1.59kHz 1 0
ROW3	15.9kHz 0.1 −45
ROW4	159kHz 0.01 −90
ROW5	1590k 0.001 −90

FIGURE 1.12: Enter the FTABLE values

From the **Analysis Setup,** select **AC Sweep** and then **D̲ecade, Point/Decade** = 10,001, **S̲tart Frequency** = 1, and **E̲nd Frequency** = 10 Meg. Press **F11** to simulate and plot the asymptotic amplitude response shown in Fig. 1.13.

The **Ftable** values are entered in magnitude/ phase domain (R_I=) or complex number domain (R_I=YES) form. The values are interpolated and converted to dB, and the phase in degrees. Frequency components outside the table range have zero dB level and impose upper and lower limits on the output. The **DELAY** attribute increases the group delay of the frequency table and is useful if an **FTABLE** part generates a noncausality warning message during a transient analysis. If such is the case, then the delay value is assigned to the DELAY attribute.

1.6   HIGH-PASS CR FILTER

The high-pass filter in Fig. 1.14 has a 1 V input signal applied using a **VSIN** generator part. Place a **Net alias** name **vout** on the output wire segment when selected (or place a **Portleft tag** instead**)** and attach voltage markers as shown.

Applying the potential the divider gives the transfer function as

$$\frac{V_2}{V_1} = \frac{R}{R + 1/j\omega C}\bigg/\frac{R}{R} = \frac{1}{1 - j/\omega CR} = \frac{1}{1 - jf_p/f}. \qquad (1.18)$$

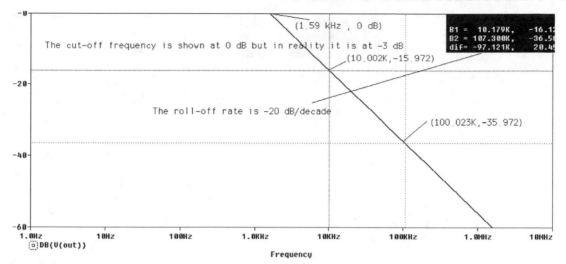

FIGURE 1.13: The FTABLE frequency response

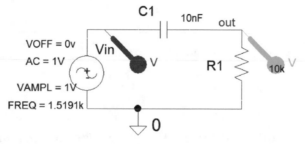

FIGURE 1.14: High-pass CR first-order filter

The cut-off frequency $\omega_p = 1/CR \Rightarrow f_p = 1/2\pi CR = 1.59$ kHz and the magnitude of the transfer function in dB is

$$|TF|_{dB} = -10\log[1 + (f_p/f)^2].\qquad(1.19)$$

At $\omega = \omega_p/10$, the magnitude of the transfer function is $|TF|_{dB} = -10\log[101] \approx -20$dB. Similarly, a decade above the cut-off frequency $f = 10f_p$ makes the transfer function magnitude $|TF|_{dB} = -10\log[1 + 1/101] \approx 0$ dB. Plot the straight-line approximate amplitude response.

1.6.1   Phase Response Measurement

An expression for the transfer function phase is $\phi = \tan^{-1}(f_p/f)$. Remember $-\tan^{-1}(-x) = \tan^{-1} x$. At $f = f_p/10$, the phase angle is $\phi = \tan^{-1}(10f/f_p) = \tan^{-1} 10 = +90°$. At $f = f_p$, the phase is $\phi = \tan^{-1}(f_p/f_p) = \tan^{-1} 1 = +45°$. At $\omega = 10\omega_p$, the phase angle is

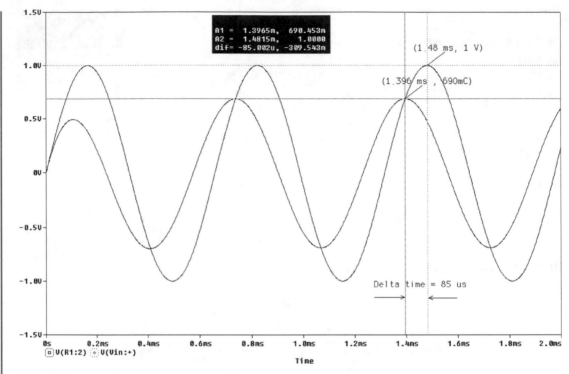

FIGURE 1.15: Measuring the phase difference

$\phi = \tan^{-1}(f_p/10f_p) = \tan^{-1}(1/10) = 0°$. The phase between input and output signals at a particular frequency is measured setting the **VSIN** generator part frequency **FREQ** to 1.519 kHz, and markers placed on the input and output. Set **Run to time** to 2 ms, and **Maximum step size** = 1 μ. Press **F11** to simulate and display the out-of-phase input and output signals as shown in Fig. 1.15. The phase difference between the two signals is measured indirectly by measuring the time difference between the signals using the two cursors. We do not measure the phase difference at the start but wait until all transient signals have died down. Convert the measured time delay to phase in degrees using the expression

$$\theta = t^*360^*\text{frequency} = 85 \text{ μs}^*360^*1.519 \text{ kHz} = 45° \qquad (1.20)$$

where the frequency is 1.519 kHz.

The high-pass filter output is across the resistance, but we may use a pair of differential voltage markers from the menu, to plot the voltage across the capacitance to give a low-pass filter response. From the **Analysis Setup,** select **AC Sweep** and then Logarithmic, **Point/Decade** = 1001, **Start Frequency** = 0.1, and **End Frequency** = 100k. To plot amplitude and phase responses, add Vdb and VP markers from the **PSpice/Markers/Advanced** menu. Press **F11**

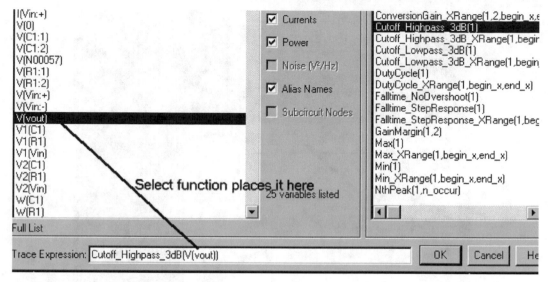

FIGURE 1.16: Using the measurement menu

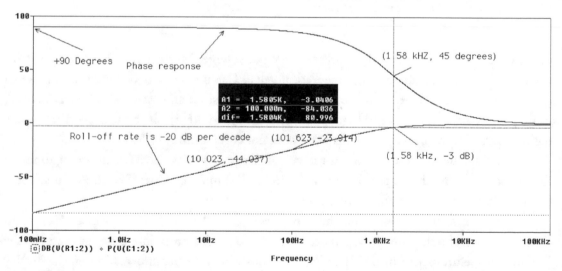

FIGURE 1.17: Phase response

to simulate. We will now measure the high pass cut-off frequency using an inbuilt function. Selecting **Trace/Evaluate measurement** produces the menu shown in Fig. 1.16.

Select **Cutoff_Highpass_3dB(1)** from the right list and substitute the output voltage variable V(vout) from the left list of variables into the function as shown (i.e., replace 1 with V(vout)) to give an accurate value for the cut-off frequency. Fig. 1.17 shows the phase response

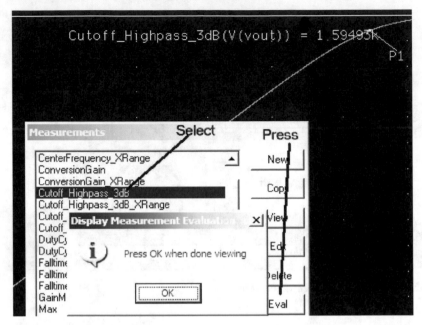

FIGURE 1.18: Evaluate measurement function

of the output voltage phase with respect to the input voltage. Measure the phase at the cut-off frequency ω_p, and compare to the theoretical value. Use two cursors to read the number of degrees per octave change in frequency. We need to consider an octave, which is a doubling in frequency, e.g., 1 kHz to 2 kHz (a decade is 1 kHz to 10 kHz), since the roll-off frequency region (transition region) is quite small and does not cover a full decade.

Another method for measuring the cut-off frequency uses the **Measurement** function that is accessed from the same **Trace** menu in Probe. The function **Cutoff High 3 dB** is selected from the list on the right panel in Fig. 1.18.

Press the **Eval button**. From the new screen, enter **V(vout)** in the **Name of Trace to search text** box should display the required information. This procedure is not as useful as the **Evaluate Measurement** function described previously because the information on the display disappears when the ok button is pressed.

1.7   GROUP DELAY

All filters produce a phase between the input and output signals (i.e., time delay between the input and output). If this phase is constant for all frequencies then is no problem. However, if each frequency in the input signal experiences different amounts of phase shift then the phase response is not linear and the output signal will be distorted. One way of representing phase

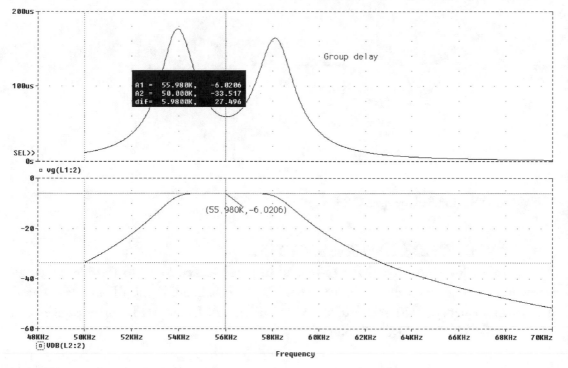

FIGURE 1.19: Group delay and bandpass response

shift change with frequency is the concept of group delay D and is defined as the derivative of the phase shift with respect to frequency.

$$D = -\frac{d\phi}{df}. \qquad (1.21)$$

Thus, the group delay is the slope of the phase response and is important for certain classes of signals such as bandpass signals. Let us consider an amplitude-modulated carrier signal applied to a bandpass filter. It is desirable that each frequency component in the signal is exposed to the same phase response; otherwise the signal will be distorted. Fig. 1.19 shows the frequency response for a sixth-order bandpass filter and the nonlinear group delay response is shown on the top. It is obvious that the group delay is not constant over the bandpass region and this could cause problems because each frequency in the signal would experience a different delay. Place a group delay marker from the **PSpice/Marker/Advanced** menu or from the **Trace Add** menu, select the **G()** operator (this was **VG()** in version 8) and substitute the variable in the brackets.

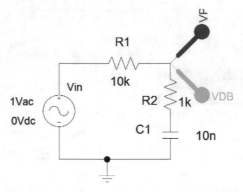

FIGURE 1.20: Modified low-pass CR filter

1.8   MODIFIED LOW-PASS FILTER

The modified low-pass filter (MLPF) is used in phase locked loops to provide a constant attenuation at high frequencies, unlike the LPF, which falls off at −20 dB for every decade increase in frequency. It also speeds up the response of the PLL compared to an uncompensated PLL. For the MLPF in Fig. 1.20, obtain:

- The amplitude and phase response,
- The two break frequencies,
- The constant attenuation factor in dB, and

What is the total phase at the first cut-off frequency?

The numerator and denominator of the transfer function are multiplied by $j\omega C$:

$$TF = \frac{V2}{V1} = \frac{R_2 + 1/j\omega C}{R_1 + R_2 + 1/j\omega C}\left(\frac{j\omega C}{j\omega C}\right) = \frac{1 + j\omega C R_2}{1 + j\omega C(R_1 + R_2)} = \frac{1 + j\omega/\omega_{p1}}{1 + j\omega/\omega_{p2}}. \quad (1.22)$$

The two cut-off frequencies are $\omega_{p1} = 1/C R_2\,\mathrm{rs}^{-1}$ and $\omega_{p2} = 1/C(R_1 + R_2)\,\mathrm{rs}^{-1}$, where $\omega_{p1} > \omega_{p2}$. Express (1.22) in magnitude and phase form as

$$|TF| = \frac{\sqrt{1 + (\omega/\omega_{p1})^2}}{\sqrt{1 + (\omega/\omega_{p2})^2}}\angle\left[\tan^{-1}\left(\frac{\omega}{\omega_{p1}}\right) - \tan^{-1}\left(\frac{\omega}{\omega_{p2}}\right)\right]. \quad (1.23)$$

From the **Analysis** menu, select **AC Sweep** and **Logarithmic, Point/Decade** = 1001, **Start Frequency** = 10, and **End Frequency** = 100 k. Press **F11** to simulate and plot the amplitude and phase responses shown in Fig. 1.21. Measure the constant attenuation $20\log[R_2/(R_1 + R_2)]$

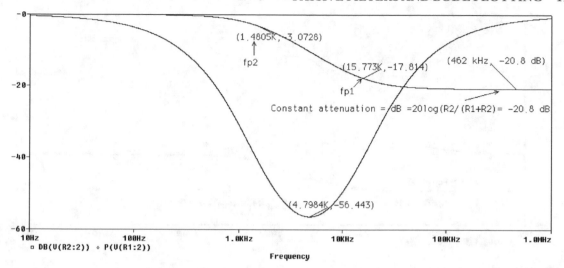

FIGURE 1.21: Modified low-pass amplitude and phase response

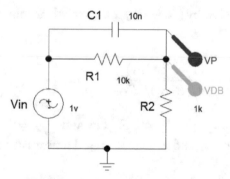

FIGURE 1.22: Modified high-pass filter

and compare to hand calculations. Calculate the phase at ω_{p1}, compare to the measured value and measure the minimum phase value and the frequency where it occurs.

1.8.1   Modified High-Pass Filter
The transfer function for the circuit in Fig. 1.22 is obtained by applying the potential divider principle.

$$TF = \frac{R_2}{R_2 + R_1/(1 + j\omega C R_1)} \times \frac{1 + j\omega C R_1}{1 + j\omega C R_1} = \frac{R_2(1 + j\omega C R_1)}{R_2(1 + j\omega C R_1) + R_1}$$

$$= \frac{R_2(1 + j\omega C R_1)}{R_1 + R_2 + j\omega C R_1 R_2}.$$

$$(1.24)$$

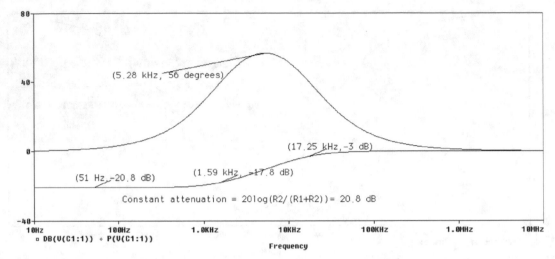

FIGURE 1.23: Modified high-pass filter phase and amplitude response

Dividing above and below by $(R_1 + R_2)$ makes the denominator real part one (A requirement for Bode plotting).

$$TF = \left(\frac{R_2}{R_1 + R_2}\right)\frac{(1 + j\omega C R_1)}{1 + j\omega C(R_2 R_1)/(R_1 + R_2)} = \frac{R_2}{R_1 + R_2}\left(\frac{1 + j\omega/\omega_{p1}}{1 + j\omega/\omega_{p2}}\right). \quad (1.25)$$

The cut-off frequencies are: $f_{p1} = 1/2\pi C R_1$ Hz and $f_{p2} = 1/2\pi C(R_1 R_2)/(R_1 + R_2)$ Hz. Thus, the TF has a zero at ω_{p1} and a pole at ω_{p2}. The transfer function in magnitude and phase is

$$|TF| = \frac{R_2}{R_1 + R_2}\frac{\sqrt{1 + (\omega/\omega_{p1})^2}}{\sqrt{1 + (\omega/\omega_{p2})^2}}\angle[\tan^{-1}(\omega/\omega_{p1}) - \tan^{-1}(\omega/\omega_{p2})]. \quad (1.26)$$

The TF magnitude part is expressed in decibels as

$$|TF| = 20\log\frac{R_2}{R_1 + R_2} + 10\log\left[1 + \left(\frac{\omega}{\omega_{p1}}\right)^2\right] - 10\log\left[1 + \left(\frac{\omega}{\omega_{p2}}\right)^2\right].$$

From the **Analysis Setup**, select **AC Sweep** and **Logarithmic, Point/Decade** = 1001, **Start Frequency** = 10, and **End Frequency** = 10meg. Press **F11** to simulate and obtain the amplitude and phase response for the modified high-pass filter (MHPF) as shown in Fig. 1.23.

Measure the constant attenuation factor $20\log(R_2/(R_1 + R_2))$ and phase at the cut-frequency ω_{p1} and compare to the values by a calculator. Measure the maximum phase value and the frequency where it occurs.

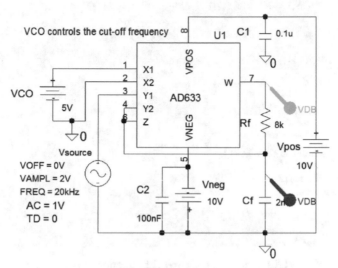

FIGURE 1.24: Low-pass filter with VCO controlling the cut-off frequency

1.8.2 Voltage-Controlled Low-Pass and High-Pass Filters

The AD633 in Fig. 1.24 is a four-quadrant multiplier IC and is configured here as a voltage-controlled low-pass filter with the −3 dB cut-off frequency controlled by the input voltage called VCO. (See ref: 3 for more information on the AD633.)

The output voltage at pin 7 is modulated by the DC input called VCO and is expressed as

$$V_{\text{out}} = \frac{VCO/10}{2\pi R_f C_f}.$$ (1.27)

The −3 dB cut-off frequency is $f_p = 1/(2\pi R_f C_f)$. From **Analysis type menu** tick **AC Sweep** with **AC Sweep Type = logarithmic, Point/Decade = 1001, Start Frequency = 10 Hz, End Frequency = 1 Meg**. The roll-off rate is −20 dB/decade, but if the frequency range over which you are measuring is quite small, then tick **Octave** (−20 dB/decade = −6 dB/octave). For example, the output across C_f has a pole at frequencies from 100 Hz to 10 kHz, for VCO voltages 100 mV to 10 V, and $R = 8$ kΩ and $C = 0.002$ µF. We sweep the DC source called **VCO,** in order to change the cut-off frequency. The **Parametric Sweep** parameter settings are shown in Fig. 1.25.

The results of sweeping VCO are shown in the multiple ac responses shown in Fig. 1.26.

A high-pass filter is achieved by inter-changing C_f and R_f.

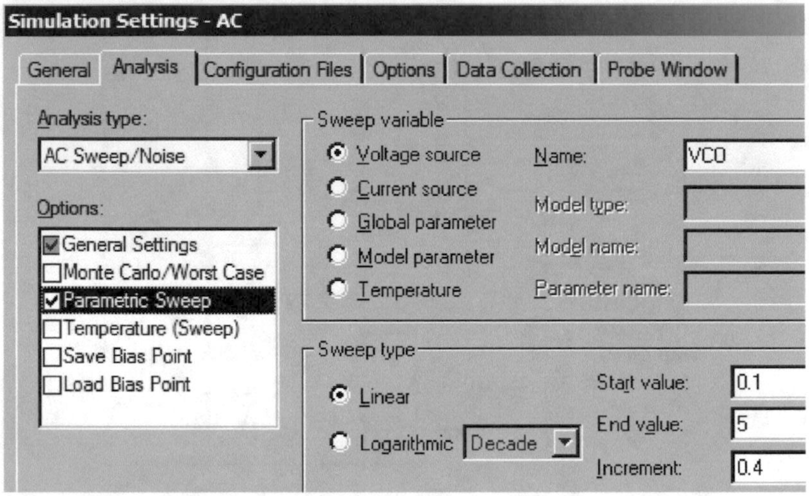

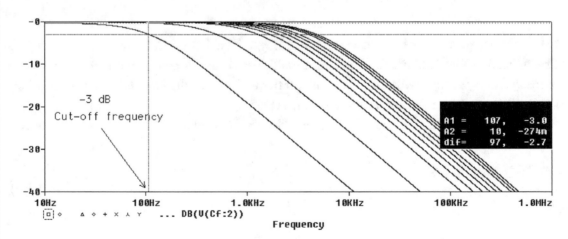

FIGURE 1.25: Sweeping VCO to change the cut-off frequency

FIGURE 1.26: LPF frequency response

1.9   NICHOLS CHART AND A LAG NETWORK

A Nichols chart investigates stability of closed-loop systems by plotting the y-axis output in dB versus the x-axis output phase. Fig. 1.27 shows not how a Laplace part can be used to plot the lay network transfer function.

From the **Analysis Setup** menu, select **AC Sweep** and **Logarithmic, Point/Decade** = 10001, **Start Frequency** = 0.001, and **End Frequency** = 100. Press **F11** to simulate and plot the output amplitude in dB. Change the x-axis to **P(V(Out))** and change the x-axis to linear and with x-axis range as shown to produce the Nichols chart Fig. 1.28.

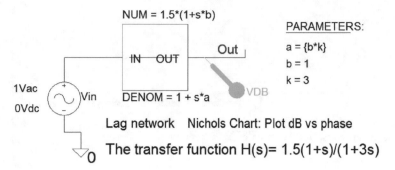

NUM = 1.5*(1+s*b)

PARAMETERS:

a = {b*k}
b = 1
k = 3

Lag network   Nichols Chart: Plot dB vs phase

The transfer function H(s)= 1.5(1+s)/(1+3s)

FIGURE 1.27: Lag network

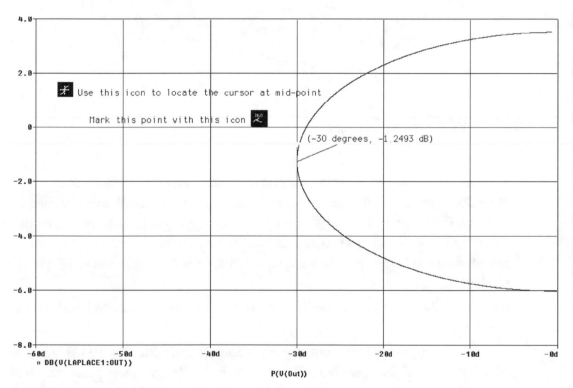

FIGURE 1.28: Nichols plot

1.10   EXERCISES

1. Use the **FTABLE** part to produce a Bode amplitude response for the MLPF. Replace the **VAC** generator with a VPULSE generator part whose parameters are: **V1** = 1, **V2** = 0, **TD** = 1 ms, **TR**=**TF** = 1 n, **PW** = 0.5 ms and **PER** = 1 ms and investigate the filtering effect on the frequency spectrum.

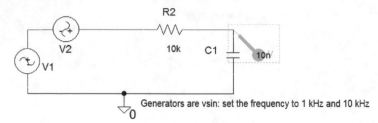

FIGURE 1.29: Measuring the second VSIN frequency

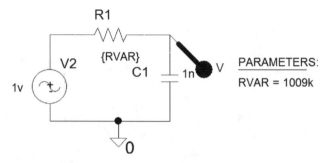

FIGURE 1.30: Varying the cut-off frequecy

2. To cement the ideas of filtering firmly in your mind, create the schematic shown in Fig. 1.29. Replace the **VAC** generators with **VSIN** generators and fill in the parameters such as **VOFF** (the offset voltage). Set up a transient analysis and observe the output. Use the **FFT** icon button to observe the frequency content and measure the relative amplitudes of each frequency in the spectrum output and compare the amplitudes to previous values.

3. Investigate the effect of varying the cut-off frequency using the **PARAM** part in Fig. 1.30.

4. Draw the high-pass CR filter investigated previously and with $R = 10$ kΩ and $C = 10$ nF, the cut-off frequency is 1.591 kHz. Set the VSIN generator to this frequency and perform a transient analysis: **Run to time** $= 2$ ms and **Maximum step size** $= 1$ μs. In Probe, plot the y-axis as the output voltage (i.e., across R) and change the x-axis to the capacitor voltage and the Lissajous plot in Fig. 1.31 should appear. Measure the maximum voltage at 0 V and the minimum value at the top of the plot and calculate the phase between the input and output voltages at f_c as

$$\arcsin\left(\frac{V_{\min}}{V_{\max}}\right) = \arcsin\left(\frac{0.707}{1}\right) = 45°. \qquad (1.28)$$

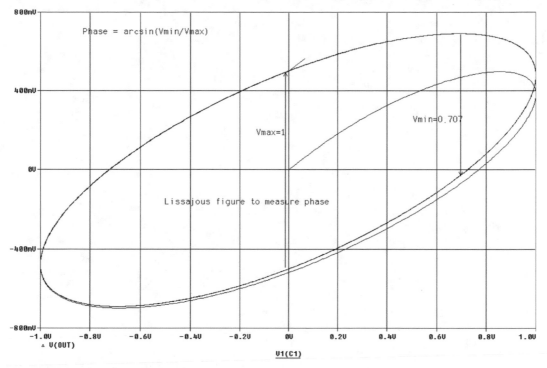

FIGURE 1.31: Lissajous figure for phase measurement

The phase is known in this example, since the frequency is set to the cut-off frequency of the network. If you are using Matlab to do your calculations, make sure you multiply the result by 180/pi, since Matlab operates in radian measure, i.e., asin (0.707/1)*180/ pi = 45°.

CHAPTER 2

Loss Functions and Active
Filter Design

2.1   APPROXIMATION LOSS FUNCTIONS

Active filters use capacitors, resistors, and op-amps (in general, inductors are not used as they tend to be large at low frequencies and also radiate magnetic fields). They also require DC power supplies and have a frequency range limited by the quality of the operational amplifier used. To design these filters, we carry out a mathematical analysis that uses approximation loss functions $A(\$)$, which when inverted produce a filter transfer function $H(\$)$, i.e.,

$$H(\$) = \frac{1}{A(\$)}. \qquad (2.1)$$

This inversion turns an all-zero function into an all-pole transfer function. Popular approximation loss functions are:

- Bessel,
- Butterworth,
- Chebychev, and
- Elliptical

2.2   BUTTERWORTH APPROXIMATING FUNCTION

A low-pass Butterworth approximation loss function, in magnitude squared form, is

$$\left| A(j\omega) \right|^2 = 1 + \left| F(j\omega) \right|^2. \qquad (2.2)$$

where the maximally-flat Butterworth characteristic function is defined as

$$\left| F(j\omega) \right| = \varepsilon(\omega/\omega_p)^n. \qquad (2.3)$$

Epsilon is a function that allows $A_{\max}$ to be adjusted to a value less than -3 dB but more about that later. Substituting (2.3) into (2.2) yields an expression for the magnitude of the loss function

$$|A(j\omega)| = \left[1 + \varepsilon^2(\omega/\omega_p)^{2n}\right]^{1/2}. \tag{2.4}$$

The loss function attenuation in the passband region $0 < \omega < \omega_p$ is called A_{max} and the attenuation in the stopband region $\omega \geq \omega_s$ is A_{min}; expressed in dB as

$$A(j\omega) = 20 \log |A(j\omega)| \text{ dB}. \tag{2.5}$$

2.2.1   The Frequency Scaling Factor and Filter Order

To obtain an expression for the frequency scaling factor ε we must substitute $\omega = \omega_p$ into (2.4).

$$A(\omega)\big|_{\omega=\omega_p} = A_{max} = 10 \log_{10}[1 + \varepsilon^2(\omega_p/\omega_p)^{2n}] \Rightarrow A_{max}$$

$$= 10 \log_{10}[1 + \varepsilon^2] \Rightarrow \varepsilon = \sqrt{10^{0.1A_{max}} - 1}. \tag{2.6}$$

For example, when A_{max} is equal to 1 dB, then the frequency correction factor ε is equal to 0.508 but when $A_{max} = 3$ dB, a popular choice, then ε is 1. The attenuation at the stopband edge frequency ω_s is A_{min},

$$A(\omega)\big|_{\omega=\omega_s} = A_{min} = 10 \log_{10}[1 + \varepsilon^2(\omega_s/\omega_p)^{2n}] \Rightarrow 0.1A_{min}$$

$$= \log_{10}[1 + \varepsilon^2(\omega_s/\omega_p)^{2n}]. \tag{2.7}$$

Taking anti-logs of both sides

$$1 + \varepsilon^2(\omega_s/\omega_p)^{2n} = 10^{0.1A_{min}} \Rightarrow (\omega_s/\omega_p)^{2n} = \frac{10^{0.1A_{min}} - 1}{\varepsilon^2}. \tag{2.8}$$

Or substituting for $\varepsilon^2 = 10^{0.1A_{max}} - 1$ from (2.6) yields

$$(\omega_s/\omega_p)^{2n} = \frac{10^{0.1A_{min}} - 1}{10^{0.1A_{max}} - 1}. \tag{2.9}$$

This yields an expression for the order of the loss function n by taking logs of both sides of (2.9) as

$$n = \frac{\log_{10}\left[\frac{10^{0.1A_{min}} - 1}{10^{0.1A_{max}} - 1}\right]}{2 \log_{10}(\omega_s/\omega_p)}. \tag{2.10}$$

We then go to the Butterworth loss function tables and select the desired loss function. We must invert and denormalize the chosen function to produce the required transfer function and at our required cut-off frequency f_p. For example, a first-order normalized Butterworth loss function is $1 + \$$, where $\$$ is the normalized complex frequency variable. Invert and denormalize

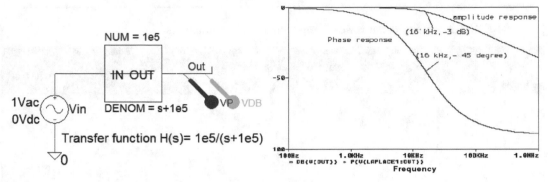

FIGURE 2.1: Laplace part to plot amplitude and phase response

this using $\$ = s/\omega_p$ to produce a TF whose cut-off frequency is ω_p, i.e.,

$$H(s) = \frac{1}{A(\$)}\Bigg|_{\$=S/\omega_p} = \frac{1}{s/\omega_p + 1} = \frac{\omega_p}{s + \omega_p}. \qquad (2.11)$$

Compare this equation (2.11) to a low-pass CR transfer function

$$TF(s) = \frac{V_o(s)}{V_{in}(s)} = \frac{1/sC}{R + 1/sC} = \frac{1}{1 + sCR} = \frac{1/CR}{s + 1/CR} = \frac{\omega_p}{s + \omega_p}. \qquad (2.12)$$

The denominator of (2.12) is a first-order Butterworth approximation loss function, with a pole at $s = -\omega_p = -1/CR$. We may plot this transfer function using a Laplace part shown in Fig. 2.1. From the **Analysis Setup,** select **AC Sweep/Logarithmic, Point/Decade** = 1000, <u>S</u>**tart Frequency** = 1, and <u>E</u>**nd Frequency** = 1000k. Press **F11** to simulate and verify that the cut-off frequency is $f_p = \frac{1}{2\pi CR} = \frac{1}{2\pi 10\times10^{-9}10^3} = 15.916$ kHz.

The attenuation at the stopband edge frequency is 20 dB and the roll-off rate from 100 kHz to 1 MHz (a decade) is −20 dB per decade.

2.3   BUTTERWORTH TABLES
Table 2.1 shows Butterworth approximation loss functions in factored form.

2.4   CHEBYCHEV APPROXIMATING FUNCTIONS
Butterworth approximation loss functions have a maximally flat passband gain with no amplitude variations (ripple). However, the attenuation in the transition region, for a given order, is not as good as the Chebychev (Pafnuty Chebychev 1821–1894) approximating loss functions. These functions have, however, ripple in the passband region called *equiripple,* where the passband gain oscillates between maximum and minimum values at a value that is determined by

TABLE 2.1: Normalized Butterworth approximation loss functions

ORDER	BUTTERWORTH APPROXIMATION LOSS FUNCTION A($) IN FACTORED FORM
1	$(\$+1)$
2	$(\$^2 + 1.414\$ + 1)$
3	$(\$+1)(\$^2 + \$ + 1)$
4	$(\$^2 + 0.765\$ + 1)(\$^2 + 1.848\$ + 1)$
5	$(\$+1)\,(\$^2 + 0.618\$ + 1)(\$^2 + 1.618\$ + 1)$
6	$(\$^2 + 0.518\$ + 1)(\$^2 + 1.414\$ + 1)(\$^2 + 1.992\$ + 1)$
7	$(\$+1)(\$^2 + 0.445\$ + 1)(\$^2 + 1.247\$ + 1)(\$^2 + 1.802\$ + 1)$
8	$(\$^2 + 0.99\$ + 1)(\$^2 + 1.111\$ + 1)(\$ + 1.669\$ + 1)(\$^2 + 1.92\$ + 1)$
9	$(\$+1)(\$^2 + 0.947\$ + 1)(\$^2 + \$ + 1)(\$^2 + 1.592\$ + 1)(\$^2 + 1.879\$ + 1)$

the choice of function (see the tables below). Chebychev polynomials are defined:

$$C_n(\omega) = \cos(n\cos^{-1}\omega) \quad \text{for} \quad -1 < \omega < 1$$
$$C_n(\omega) = \cosh(n\cosh^{-1}\omega) \quad \text{for} \quad \omega > 1 \tag{2.13}$$

where $\cosh(\omega) = 1/2(e^{\omega} + e^{-\omega})$. After some manipulation, equation (2.13) results in the following loss function whose magnitude is

$$\left|A(j\omega)\right|^2 = 1 + \varepsilon^2 C_n^2(\omega). \tag{2.14}$$

$A(j\omega)$ is a low-pass approximating function and ε defines the ripple amplitude (not to be confused with ε, the frequency-correction factor from the Butterworth design). Expressing (2.14) in dB:

$$\left|A(j\omega)\right|_{dB} = 20\log\left|A(j\omega)\right| = 20\log\{1 + \varepsilon^2 C_n^2(\omega)\}^{1/2} = 10\log\{1 + \varepsilon^2 C_n^2(\omega)\}. \tag{2.15}$$

For $C_n(\omega) \gg 1$

$$\left|A(j\omega)\right|_{dB} = 20\log\varepsilon + 20\log C_n(\omega) \tag{2.16}$$

where $C_n(\omega) \cong 2^{n-1}\omega^n$ for $\omega > 1$, so (2.16) is

$$\left|A(j\omega)\right|_{dB} = 20\log\varepsilon + 20\log 2^{n-1}\omega^n = 20\log\varepsilon + 20(n-1)\log 2 + 20\log\omega^n. \tag{2.17}$$

Comparing (2.17) to a Butterworth second-order loss function, we see that there is an extra term $20(n-1)\log 2$ equal to $6.02(n-1)$ dB, thus giving the second-order Chebychev function

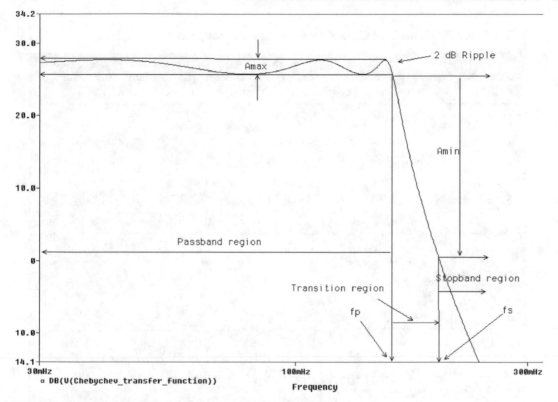

FIGURE 2.2: Chebychev loss functions and edge frequencies

an extra 6 dB fall-off in the transition region. Fig. 2.2 defines the filter parameters such as attenuation and edge frequencies.

2.4.1 Chebychev Tables
Chebychev approximation loss functions, with passband ripple 0.5 dB to 2 dB, are shown in Table 2.2.

2.4.2 Ripple Factor and Chebychev Order
The ripple factor is expressed in terms of the passband edge frequency attenuation factor, A_{max}, as

$$\varepsilon = \sqrt{10^{0.1A_{max}} - 1}. \qquad (2.18)$$

This is the same expression as derived for the Butterworth loss function but for different reasons. The specification for a filter includes the passband and stopband edge frequencies, f_p and f_s, and the attenuation at these frequencies is A_{max} and A_{min} (the passband and stopband

TABLE 2.2: Normalized Chebychev approximation loss functions

ORDER	RIPPLE = 0.1 dB ($\varepsilon = 0.1526$)
N	Chebychev approximation loss function A($) in factored form
1	($ + 6.552)
2	($^2 + 2.372$ + 3.314)
3	($ + 0.969)($^2 + 0.969$ + 1.69)
4	($^2 + 0.528$ + 1.33)($^2 + 1.275$ + 0.623)
5	($ + 0.539) ($^2 + 0.333$ + 1.195)($^2 + 0.872$ + 0.636)
6	($^2 + 0.229$ + 1.129)($^2 + 0.627$ + 0.696) ($^2 + 0.856$ + 0.263)
	Ripple = 0.5 dB ($\varepsilon = 0.349$)
1	($ + 2.863)
2	($^2 + 1.426$ + 1.516)
3	($ + 0.626)($^2 + 0.626$ + 1.142)
4	($^2 + 0.351$ + 1.064)($^2 + 0.847$ + 0.356)
5	($ + 0.362) ($^2 + 0.224$ + 1.036)($^2 + 0.586$ + 0.477)
6	($^2 + 0.155$ + 1.023)($^2 + 0.424$ + 0.59) ($^2 + 0.58$ + 0.157)
	Ripple = 1.0 dB ($\varepsilon = 0.509$)
1	($ + 1.965)
2	($^2 + 1.098$ + 1.103)
3	($ + 0.494)($^2 + 0.494$ + 0.994)
4	($^2 + 0.279$ + 0.987)($^2 + 0.674$ + 0.279)
5	($ + 0.289) ($^2 + 0.179$ + 0.988)($^2 + 0.468$ + 0.429)
6	($^2 + 0.124$ + 0.991)($^2 + 0.34$ + 0.558) ($^2 + 0.464$ + 0.125)
	Ripple = 2.0 dB ($\varepsilon = 0.765$)
1	($ + 1.308)
2	($^2 + 0.804$ + 0.823)
3	($ + 0.369)($^2 + 0.369$ + 0.886)
4	($^2 + 0.210$ + 0.929)($^2 + 0.506$ + 0.222)
5	($ + 0.218) ($^2 + 0.135$ + 0.952)($^2 + 0.353$ + 0.393)
6	($^2 + 0.094$ + 0.966)($^2 + 0.257$ + 0.533) ($^2 + 0.351$ + 0.1)

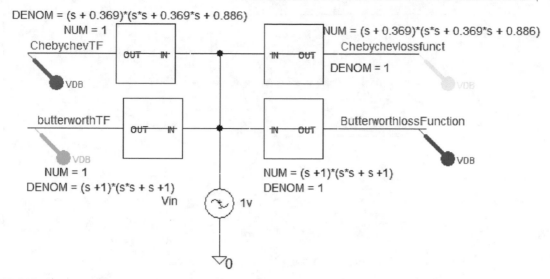

FIGURE 2.3: Laplace part for plotting transfer functions

attenuation figures respectively). At $\omega = \omega_s$, the attenuation is

$$A(\omega)\big|_{\omega=\omega_s} = A_{\min} = 10\log[1 + \varepsilon^2 C_n^2(\omega_s/\omega_p)] \Rightarrow 0.1 A_{\min}$$

$$= \log[1 + \varepsilon^2 C_n^2(\omega_s/\omega_p)]. \qquad (2.19)$$

Using this result and the ripple factor, it can be shown that the loss function order is determined as

$$n = \frac{\cosh^{-1}\left[\frac{10^{0.1 A_{\min}} - 1}{10^{0.1 A_{\max}} - 1}\right]^{1/2}}{\cosh^{-1}(\Omega_s)}. \qquad (2.20)$$

2.4.3   Plotting Chebychev and Butterworth Function

We will use **Laplace** parts in Fig. 2.3 to plot and compare loss and transfer functions. A third-order Chebychev loss function **($ + 0.369)*($*$ + 0.369*$ + 0.886)** is entered in the **DENOM** of the **Laplace** part and in the **NUM** part of a second Laplace part. This will display a Chebychev loss function and transfer function. Similarly, enter the second-order Butterworth loss function **($*$ + 1.414*$ + 1)** in the **NUM** entry of the last **Laplace** part.

From the **Analysis Setup,** select **AC Sweep, Logarithmic, Point/Decade** = 1001, **Start Frequency** = 0.01, and **End Frequency** = 10. Press **F11** to simulate. The frequency response of a Chebychev loss function and its inverse are plotted in Fig. 2.4, and shows a 2 dB ripple in the passband region. A Chebychev transfer function has a steeper roll-off rate compared to

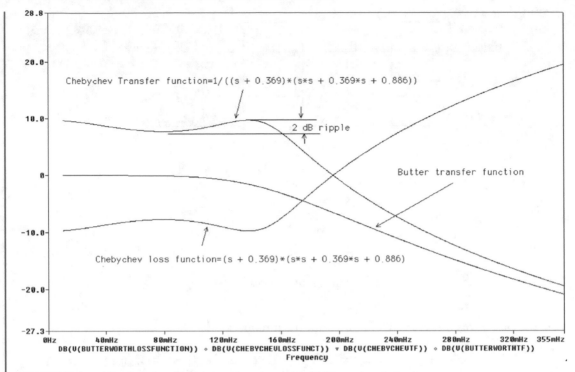

FIGURE 2.4: Third-order Chebychev function with 2 dB passband ripple

a Butterworth function but has ripples in the passband that could cause problems in certain audio and video applications.

Compare the Butterworth and Chebychev functions in Fig. 2.4. A Chebychev loss function rises at a much greater rate and hence produces a filter response that falls off at a greater rate compared to Butterworth loss functions. The phase response is important when considering complex signals such as a modulated carrier, or data signals. In this situation, it is important that the phase response is linear so that each frequency component in the signal has equal time delays otherwise the filtered signal is distorted.

2.5   AMPLITUDE RESPONSE FOR A FIRST-ORDER ACTIVE LOW-PASS FILTER

Using the **Net Alias** icon, name each power supply pin of the operational amplifier with a unique name such as −Vcc and +Vcc and do the same with the DC supplies wires. This simplifies a schematic layout containing a number of integrated circuits by connecting wire segments with the same name together. The transfer function (the voltage gain) for the low-pass active filter

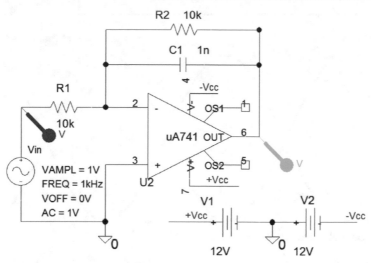

FIGURE 2.5: First-order active filter

in Fig. 2.5 is

$$A_v = \frac{V_{\text{out}}}{V_{\text{in}}} = \frac{Z_2}{Z_1} = \frac{R_2}{1 + j\omega C R_2}/R_1. \tag{2.21}$$

The gain in the magnitude form is

$$|A_v| = \frac{R_2}{R_1}\frac{1}{[1 + (2\pi f C R_2)^2]^{0.5}} = \frac{R_2}{R_1}\frac{1}{[1 + (\omega/\omega_p)^2]^{0.5}}. \tag{2.22}$$

The cut-off frequency is $\omega_p = 1/C R_2$ rs^{-1}. Set **Run to time** to 2 ms and **Maximum step size** to 400 ns. Press **F11** to simulate and display the antiphase input and output signals shown in Fig. 2.6.

Change the capacitance to 10 nF and observe the effect on the waveforms.

Change the marker to VdB. From the **Analysis Setup** menu, select **AC Sweep/Decade, Point/Decade** = 1001, **Start Frequency** = 10, and **End Frequency** = 100k. Simulate with **F11** to produce the frequency response for $R_2 = 10$ kΩ in Fig. 2.7 and resimulate for $R_2 = 20$ kΩ.

2.6   EXERCISES

(1) Investigate the **ELAPLACE** part for plotting the step response of a second-order Butterworth transfer function shown in Fig. 2.8. Substitute **VAC** for **VPWL** and plot the frequency response.

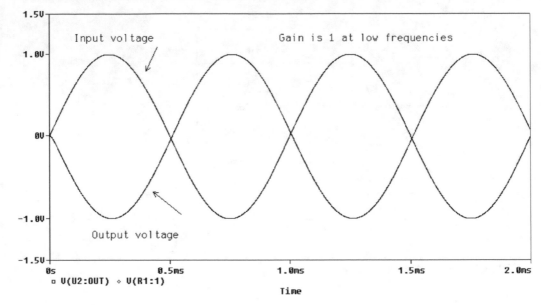

FIGURE 2.6: Waveforms for R1 = R2 = 10 kΩ, C = 1 nF

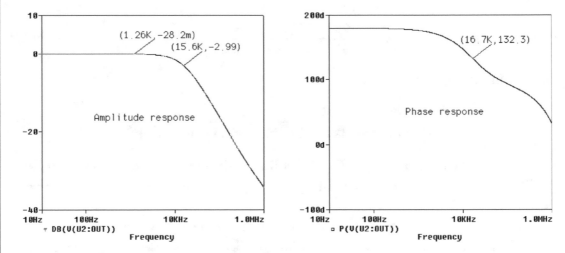

FIGURE 2.7: Frequency response for R2 = 10 kΩ

(2) Show how the order of a filter may be obtained using the expression for $A(\omega)$ and the following terms:

- Maximum passband attenuation $A_{max} = 3$ dB,
- Minimum stopband attenuation $A_{min} = 28$ dB,
- Passband edge frequency $\omega_p = 1$ krs^{-1}, and
- Stopband edge frequency $\omega_s = 10$ krs^{-1}.

CHAPTER 3

Voltage-Controlled Voltage Source Active Filters

3.1 FILTER TYPES

The procedure for designing active filters to meet a specification is shown in Fig. 3.1. The first four blocks explain how a transfer function is obtained using a specification to select a suitable loss function. The next three blocks show how a second transfer function is obtained for a second-order Sallen and Key (or an infinite gain multiple feedback) active filter circuit to implement the first TF derived from loss function analysis. The last block explains how the comparison between the two transfer function coefficients gives component values for the active filter.

Many circuit configurations exist for implementing transfer functions from loss function analysis. The following circuit configurations are used in the next few chapters:

- Sallen and Key voltage controlled voltage source (VCVS),
- Multiple feedback, infinite gain multiple feedback (IGMF),
- Biquadratic and state-variable types

We may achieve up to tenth-order filters by cascading first- and second-order types stages.

However, identical stages are rarely cascaded as it produces a poor frequency response at the cut-off frequency as shown in Fig. 3.2. For example, cascading two identical second-order filters produces a fourth-order filter with the attenuation at the cut-off frequency of −6 dB, and not −3 dB. A better technique is to design each stage with different Q factors. We will investigate voltage-controlled voltage source (VCVS) and infinite gain multiple feedback (IGMF) filter types. These have high input and low output impedances and so may be cascaded without any loading problems. We can produce higher-order filters without the need for buffering between each stage. These two types have high Q-factor sensitivities which is a measure of the effect on the response when circuit component values deviate from their stated manufactured value. This is called tolerance and is investigated using the **Monte Carlo/Worst case** facility in Probe.

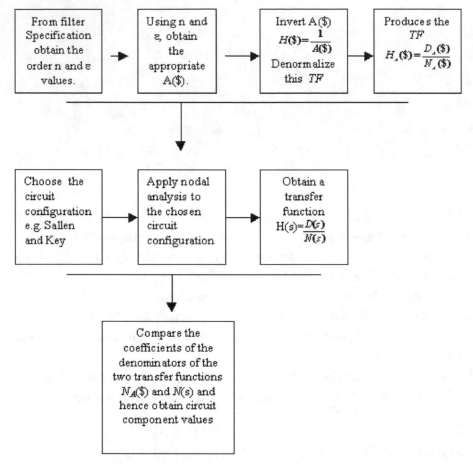

FIGURE 3.1: Active filter design procedure

Nodal analysis obtains the transfer function for each configuration and we can then compare to the transfer function from loss function analysis. Circuit values are calculated by comparing the denominator coefficients of the two transfer functions. For example, a fifth-order loss function in factored form comprises two cascaded second-order functions and a first-order function. The standard form for first- and second-order transfer functions yields expressions for the Q-factor and the passband frequency, ω_p. To reduce the degrees of freedom in the choice of component values, it is useful to use equal-component values. However, if we use a unity gain operational amplifier configuration then we cannot use equal-component values to set the damping (inverse of the Q-factor) to a particular value and require nonequal value capacitors.

The cascaded stages should be connected so that the *first stage has the lowest Q-factor* and then in each following stage should have ascending *order of Q-factors*. Consider the following

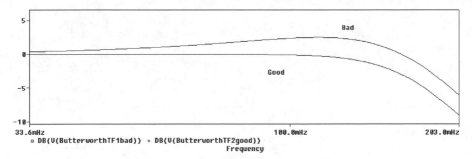

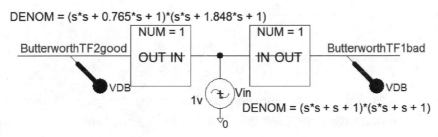

FIGURE 3.2: Cascaded sections

third- and fourth-order cascaded transfer functions in factored form:

$$H_1(s) = \left[\frac{1.602s}{s^2 + 4807}\right]\left[\frac{s^2}{s^2 + 1644s + 7.878 \times 10^6}\right] \tag{3.1}$$

$$H_2(s) = \left(\frac{28,727}{s^2 + 239.7s + 28,727}\right)\left(\frac{28,727}{s^2 + 124.7s + 28,727}\right). \tag{3.2}$$

Each stage has a different damping factor (i.e., 237.7 and 124.7). The first stage should have the lowest Q-factor value which is the same as saying the highest damping factor (in this example it is 239.7, followed by the next stage with damping equal to 124.7). The passband frequency $\omega_0^2 = 28,727$ rs^{-1} is the same for both stages because it uses a Butterworth loss function but this is not true when you use a Chebychev loss function.

3.2   VOLTAGE-CONTROLLED VOLTAGE SOURCE FILTERS (VCVS)

To simplify the analysis, we assume ideal operational amplifiers with the following characteristics:

- Infinite input impedance,
- Zero output impedance, and
- Infinite gain

Load the active first-order filter circuit shown in Fig. 3.3.

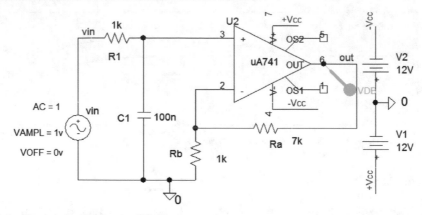

FIGURE 3.3: First-order low-pass CR filter

The low-pass filter transfer function is obtained by applying the potential divider technique. The noninverting amplifier gain is $k = 1 + Ra/Rb$, hence the complete transfer function is

$$H(s) = (1 + Ra/Rb)\left[\frac{1/CR}{s + 1/CR}\right].$$ (3.3)

The cut-off frequency is $1/(2\pi CR) = 1.6$ kHz. Substituting values into (3.3) gives the gain as

$$H(s) = (1 + 7k/1k)\left[\frac{1/10^{-9}10^3}{s + 1/10^{-9}10^3}\right] = 8\left[\frac{10^6}{s + 10^6}\right].$$ (3.4)

The passband gain ($s = 0$) in dB is $20\log(8) = 18$ dB. The frequency response is shown in Fig. 3.4 where the -3 dB frequency is read where the gain is 3 db down on the passband gain, i.e., at 15 dB and is 1.6 kHz.

We will shortly need an expression for the voltage v at the junction of the resistor and capacitor and is given as

$$\frac{E_0}{v} = k\frac{1/CR}{s + 1/CR} = k\frac{1}{1 + sCR} \Rightarrow v = \frac{E_0}{k}\frac{(sCR + 1)}{1}.$$ (3.5)

3.2.1   Sallen and Key Low-Pass Active Filter

The transfer function for the Sallen and Key second-order active low-pass filter circuit in Fig. 3.5 is derived by applying nodal analysis at the junction of R_1 and R_2, and C_1.

The sum of the currents at node v is zero (Kirchhoff's second law). However, we may write the nodal equation using the format discussed in Book 1 [ref: 1 Appendix A].

V(self-admittances at node) $+$ Vin(transfer admittance between nodes) $+$ $Vout$(transfer admittance between nodes) $= 0$. To simplify the analysis let the resistors and capacitors have

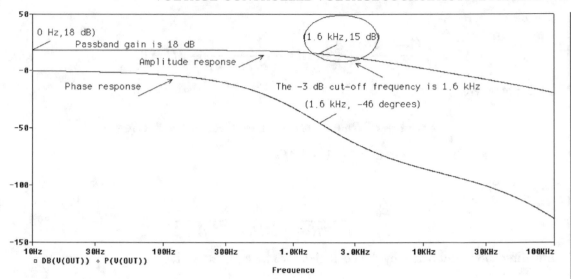

FIGURE 3.4: Amplitude response

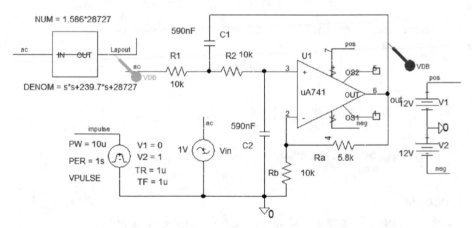

FIGURE 3.5: Sallen and Key active second-order LPF

equal values so:

$$v\left[\frac{1}{R} + \frac{1}{R} + sC\right] - \frac{E_i}{R} - \frac{E_0}{kR} - sCE_0 = 0. \qquad (3.6)$$

Substitute for v from (3.5) into (3.6) and rearrange

$$E_0\frac{(1 + sCR)}{k}\left[\frac{2}{R} + sC\right] - \frac{E_i}{R} - \frac{E_0}{kR} - sCE_0 = 0. \qquad (3.7)$$

Divide across by E_0

$$\frac{(1 + sCR)}{k}\left[\frac{2}{R} + sC\right] - \frac{E_i}{E_0 R} - \frac{1}{kR} - sC = 0. \tag{3.8}$$

Multiply by kR

$$\frac{E_i}{E_0}k = (1 + sCR)(1 + sCR) - 1 - sCRk = 1 + 3sCR + 1$$
$$+ s^2 C^2 R^2 - 1 - sCRk. \tag{3.9}$$

Invert (3.9)

$$\frac{E_0}{E_i} = \frac{k}{s^2 C^2 R^2 + s(3CR - CRk) + 1}. \tag{3.10}$$

Dividing above and below by the coefficient of s^2, i.e., $C^2 R^2$ yields

$$\frac{E_0}{E_i} = \frac{k\frac{1}{C^2 R^2}}{s^2 + s\frac{(3-k)}{CR} + \frac{1}{C^2 R^2}}. \tag{3.11}$$

Equating the denominator coefficients to the standard form yields

$$\omega_0 = \frac{1}{CR} \tag{3.12}$$

$$\frac{\omega_0}{Q} = \frac{(3-k)}{CR} \Rightarrow Q = \frac{1}{3-k}. \tag{3.13}$$

The passband gain is obtained by letting all s terms $= 0$, i.e., $G(0) = k$.

3.2.2 Example

An active low-pass filter is required to meet the following specification:

- The maximum passband loss $A_{\max} = 0.5$ dB
- The minimum stopband loss $A_{\min} = 12$ dB
- The passband edge frequency $\omega_p = 100$ rs^{-1}($f_p = \omega_p/2\pi = 15.9$ Hz)
- The stopband edge frequency $\omega_s = 400$ rs^{-1}

Obtain a Butterworth transfer function and components for a Sallen and Key active filter circuit to meet the above specification. The filter order is calculated as

$$n = \frac{\log_{10}\left[\frac{10^{0.1 A_{\min}} - 1}{10^{0.1 A_{\max}} - 1}\right]}{2\log_{10}(\omega_s/\omega_p)} = \frac{\log_{10}\left[\frac{10^{0.1 \times 12} - 1}{10^{0.1 \times 0.5} - 1}\right]}{2\log_{10}(400/100)} = 1.7 \approx 2. \tag{3.14}$$

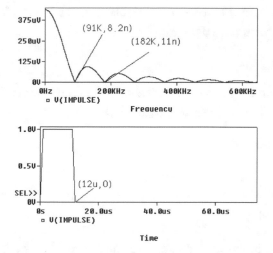
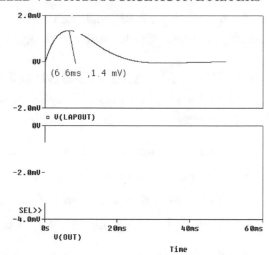

FIGURE 3.7: The input impulse function

source library with parameters:

- DC $= 0$ (offset)
- AC $= 1$ (amplitude)
- V1 $= 0$ (space value)
- V2 $= 1$ (mark value)
- TD $= 0$ s (delay)
- TR $= 1$ μs (rise time)
- TF $= 1$ μs (Fall time)
- PW $= 10$ μs (pulse width----make it short to mimic an ideal impulse)
- PER $= 1$ s (period-----make it very large to create the illusion of only one impulse)

Change *Ra* to 19 kΩ to decrease the damping and observe the decaying oscillatory output. The transient setup parameters are: **Output File Options/Print values in the output file** $= 20$ ns, **Run to time** $= 0.05$ s and **Maximum step size** $= 1$ μs. Press **F11** to produce the impulse in Fig. 3.7.

The impulse spectrum is obtained by selecting the **FFT** icon. The impulse spectrum is a sinc-shaped function sinx/x but PSpice displays absolute values only. Measure the frequency at the minimum points and determine the relationship between this frequency and the pulse width. You may display both the pulse and its spectrum (time and frequency domain) simultaneously, by selecting **Plot/Unsynchronize X-Axis** from the Probe screen. Separate the two displayed signals by pressing **alt PP**, and then apply **ctrl X and ctrl V** to a variable at the bottom. A decaying sine wave should be observed together with the 10 µs impulse.

3.3   FILTER AMPLITUDE RESPONSE

The ABM **Laplace** part allows us to plot the frequency response for transfer functions derived from loss functions. Consider the specification for an active low-pass filter is:

- The maximum passband loss $A_{max} = 1$ dB
- The minimum stopband loss $A_{min} = 20$ dB
- The passband edge frequency $f_p = 3.4$ kHz
- The stopband edge frequency $f_s = 6.8$ kHz

Determine the filter order, select a suitable loss function and calculate the loss in dB at the stopband edge frequency.

3.3.1   Solution

The filter order is calculated:

$$n = \frac{\log_{10}\left[\frac{(10^{0.1 A_{min}}-1)}{(10^{0.1 A_{max}}-1)}\right]}{2\log_{10}(\omega_s/\omega_p)} = \frac{\log_{10}\left[\frac{(10^{0.1\times 20}-1)}{(10^{0.1\times 1}-1)}\right]}{2\log_{10}(6800/3400)} = \frac{\log_{10}(99/0.2589)}{2\log_{10}(2)} = \frac{2.58}{0.602} = 4.29 \approx 5.$$

$$(3.27)$$

Select the approximation loss function from the loss function tables:

$$A(\$) = (\$+1)(\$^2+0.618\$+1)(\$^2+1.616\$+1). \tag{3.28}$$

Inverting (3.28) yields the transfer function

$$H(\$) = \frac{1}{A(\$)} = \frac{1}{(\$+1)(\$^2+0.618\$+1)(\$^2+1.616\$+1)}. \tag{3.29}$$

Denormalize the loss function by replacing $\$$ with $s\varepsilon^{1/n}/\omega_p$. Epsilon, the frequency shift parameter, is calculated as $\varepsilon = \sqrt{(10^{0.1.A_{max}}-1)} = \sqrt{(10^{0.1\times 1}-1)} = 0.508$.

$$\$ = s\frac{\varepsilon^{1/n}}{\omega_p} = s\frac{0.508^{1/5}}{2\pi \times 3400} = \frac{s}{24456} \tag{3.30}$$

$$H(s)|_{\$=s/24456} = \frac{1}{\left[\frac{s}{24456}+1\right]\left[\left(\frac{s}{24456}\right)^2+0.618\frac{s}{24456}+1\right]\left[\left(\frac{s}{24456}\right)^2+1.616\frac{s}{24456}+1\right]} \tag{3.31}$$

$$H(s) = \frac{(24456)^5}{[s+24456][s^2+15114s+24456^2][(s^2+39521s+24456^2]}. \tag{3.32}$$

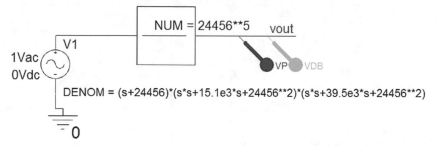

FIGURE 3.8: The ABM Laplace part

Select the **Laplace** part and enter 8.7484e21 in the **NUM**erator and enter the denominator in the **DENOM** box.

$$H(s) = \frac{8.7484\text{e}21}{(s + 24456)*(s*s + 15114*s + 24456**2)*(s*s + 39530*s + 24456**2)}. \quad (3.33)$$

Figure 3.8 shows the transfer function in the **ABM Laplace** part with the output wire segment called **vout** and with a dB marker attached. To square a function, enter $s*s$, or $s**2$ but you may also use **PWRS(s, 2)** as an alternative. This is useful when large powers of s are involved, i.e., **PWRS(s, 6)** for s^6.

Always use round brackets (square brackets are not allowed) to ensure that the total denominator is divided into the numerator. Be careful about multiplication and always use * between brackets, variables and numbers, e.g., $(s + 2)*(s + 1)$, and $15.91\text{e}3*s$. The attenuation at the stopband edge frequency is

$$A(s) = 10 \log_{10} \left[1 + \varepsilon^2 \left(\frac{\omega}{\omega_p} \right)^{2n} \right]. \quad (3.34)$$

The attenuation for $n = 5$

$$A(s) = 10 \log_{10}[1 + 0.259(2)^{10}] = 24.5 \text{ dB}. \quad (3.35)$$

We must express the transfer function in factored form because we wish to implement each factor using individually designed stages. From the **Analysis Setup,** select **AC Sweep,** and **Logarithmic, Point/Decade** = 1001, **Start Frequency** = 10, and **End Frequency** = 100k. Press **F11** to simulate and plot the fifth-order Butterworth transfer function amplitude response

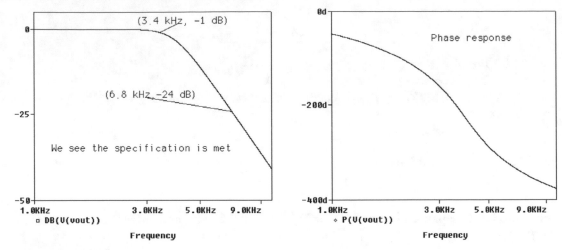

FIGURE 3.9: Amplitude response

as shown in Fig. 3.9. You may use the **Evaluate Measurement Function** to check if the specification is met.

3.4 BUTTERWORTH THIRD ORDER ACTIVE HIGH-PASS FILTER

Obtain a Butterworth high-pass transfer function to meet the following specification:

- The maximum passband loss $A_{max} = 3$ dB
- The minimum stopband loss $A_{min} = 28$ dB
- The passband edge frequency $\omega_p = 6000$ rs^{-1}
- The stopband edge frequency $\omega_s = 2000$ rs^{-1}

Calculate component values for the third-order active filter circuit in Fig. 3.10 to implement this transfer function.

Let $R = 20$ kΩ. The frequency correction factor is one since A_{max} is -3 dB, or calculated as

$$\varepsilon = \sqrt{(10^{0.1 \times A_{max}} - 1)} = \sqrt{(10^{0.1 \times 3} - 1)} = 0.9976 \approx 1. \qquad (3.36)$$

The filter order is calculated as

$$n = \frac{\log_{10}\left[\frac{(10^{0.1 A_{min}} - 1)}{(10^{0.1 A_{max}} - 1)}\right]}{2\log_{10}(\omega_s / \omega_p)} = \frac{\log_{10}\left[\frac{(10^{0.1 \times 28} - 1)}{(10^{0.1 \times 3} - 1)}\right]}{2\log_{10}(6000/2000)} = 2.93 \approx 3. \qquad (3.37)$$

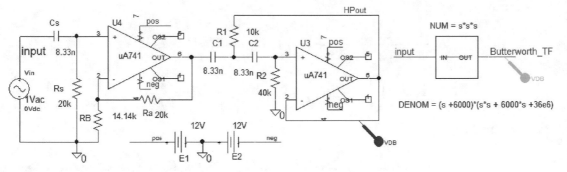

FIGURE 3.10: Third-order active filter

Select the third-order normalized loss function $A(\$) = (\$ + 1)(\$^2 + \$ + 1)$ from the Butterworth tables. This is denormalized and transformed from a high-pass to a low-pass function by replacing $\$$ with ω_p/s. Since the passband attenuation is 3 dB, then $\varepsilon = 1$.

$$A(\$)|_{\$= 6000/s} = A(s) = \left(\frac{6000}{s} + 1\right)\left(\frac{6000^2}{s^2} + \frac{6000}{s} + 1\right). \qquad (3.38)$$

The transfer function is

$$H(s) = \frac{1}{A(s)} = \frac{1}{\left(\frac{6000}{s} + 1\right)\left(\frac{6000^2}{s^2} + \frac{6000}{s} + 1\right)} = \frac{s^3}{(s + 6000)(s^2 + 6000s + 36 \times 10^6)}. \qquad (3.39)$$

The first-order transfer function is

$$\frac{E_2}{E_1} = \frac{R}{1/sC + R} = \frac{s}{s + 1/CR} = \frac{s}{s + \omega_p} = \frac{s}{s + 6000}. \qquad (3.40)$$

Substituting values for $R = 20$ kΩ and ω_p gives C as

$$C = \frac{1}{6 \times 10^3 \times 20 \times 10^4} = 8.33 \text{ nF.} \qquad (3.41)$$

3.4.1   The second-order transfer function

Apply nodal analysis to the node marked v,

$$v(1/R_1 + 2sC) - E_3sC - E_3/R_1 = E_2sC. \qquad (3.42)$$

From the potential divider rule, we may write the relationship between v and E_3 as

$$v\left(\frac{R_2}{1/sC + R_2}\right) = E_3 \Rightarrow v = E_3\frac{(1 + sCR_2)}{sCR_2}. \qquad (3.43)$$

Substituting for v yields

$$E_3 \frac{(1+sCR_2)}{sCR_2}\left(\frac{1}{R_1}+2sC\right) - \frac{E_3}{R_1} - E_3 sC = E_2 sC \qquad (3.44)$$

$$(1+sCR_2)\left(\frac{1}{R_1}+2sC\right) - \frac{sCR_2}{R_1} - s^2C^2R_2 = \frac{E_2}{E_3}s^2C^2R_2. \qquad (3.45)$$

Multiply by R_1

$$(1+sCR_2)(1+2sCR_1) - sCR_2 - s^2C^2R_2R_1 = \frac{E_2}{E_3}s^2C^2R_1R_2 \qquad (3.46)$$

$$1+2sCR_1+sCR_2+2s^2C^2R_1R_2 - sCR_2 - s^2C^2R_2R_1 = \frac{E_2}{E_3}s^2C^2R_1R_2. \qquad (3.47)$$

Substitute values into the transfer function

$$\frac{E_3}{E_2} = \frac{s^2C^2R_1R_2}{s^2C^2R_1R_2+s2CR_1+1} = \frac{s^2}{s^2+s\frac{2}{CR_2}+\frac{1}{R_1R_2C^2}} = \frac{s^2}{s^2+6000s+6000^2}. \qquad (3.48)$$

Comparing the s coefficient terms and substituting for C yields

$$\frac{2}{R_2C} = 6000 \Rightarrow R_2 = \frac{2}{8.33\times10^{-9}\times6000} = \frac{2}{8.33\times10^{-6}\times6} = \frac{10^6}{24.99} = 40\text{ k}\Omega. \quad (3.49)$$

Equate the non-s terms to obtain R_1,

$$\frac{1}{R_1R_2C^2} = 6000^2 \Rightarrow R_1 = \frac{1}{(8.33\times10^{-9})^2\times36\times10^6\times4\times10^4}$$

$$= \frac{1}{9992\times10^{-8}} \approx 10\text{ k}\Omega. \qquad (3.50)$$

From **Analysis Setup/AC Sweep** menu, set **Logarithmic, Point/Decade** = 1001, **S̲tart Frequency** = 10, and **E̲nd Frequency** = 10k. Press **F11** to simulate and, using the cursors, check the specification is met from the frequency response in Fig. 3.11.

Replace the unrealistic component values with nearest preferred values and simulate again.

3.5   EXERCISES

(1) This exercise is partly completed but correct value need to be calculated for the third-order Sallen and Key high-pass active filter shown in Fig. 3.12, to meet the following specification:

- Passband edge frequency $\omega_p = 3000$ rs^{-1} (477 Hz)
- Stopband edge frequency $\omega_s = 1000$ rs^{-1} (159 Hz)

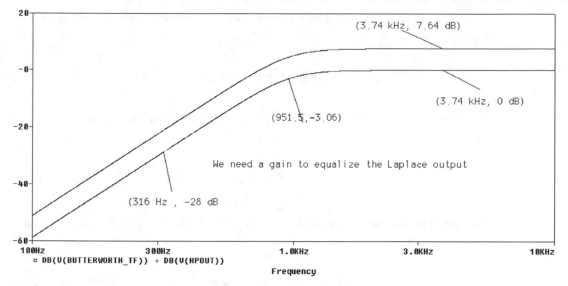

FIGURE 3.11: Response for the third-order S & K HP active filter

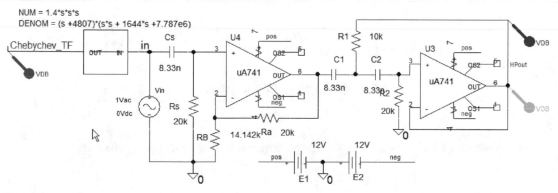

FIGURE 3.12: Chebychev Sallen and Key active third-order HPF

- Maximum passband attenuation $A_{max} = 0.5$ dB
- Minimum stopband attenuation $A_{min} = 22$ dB

Determine the Chebychev transfer function with 0.5 dB ripple and compare the components given to those calculated from the transfer function.

The filter order is determined from the following expression:

$$n = \frac{\cosh^{-1}\left[\dfrac{10^{0.1A_{min}}-1}{10^{0.1A_{max}}-1}\right]^{1/2}}{\cosh^{-1}(\omega_p/\omega_s)} = \frac{\cosh^{-1}\left[\dfrac{10^{0.1\times22}-1}{10^{0.1\times0.5}-1}\right]^{1/2}}{\cosh^{-1}(3/1)} = 2.42. \qquad (3.51)$$

From the Chebychev tables, select a third-order approximation loss function with 0.5 dB ripple.

$$A(\$) = (\$ + 0.626)(\$^2 + 0.626\$ + 1.142). \tag{3.52}$$

Convert this low-pass loss function to a high-pass loss function by making the following substitution:

$$\$ = \frac{\omega_p}{s}. \tag{3.53}$$

$\$$ is the normalized complex frequency variable and s is the complex frequency variable. The loss function is

$$A(\$)\Big|_{\$=\frac{3000}{s}} = A(s) = \left[\frac{3000}{s} + 0.626\right]\left[\left(\frac{3000}{s}\right)^2 + 0.626\left(\frac{3000}{s}\right) + 1.142\right] \tag{3.54}$$

$$A(s) = \left[\frac{3000 + 0.626s}{s}\right]\left[\frac{9 \times 10^6}{s^2} + \frac{1879}{s} + 1.142\right]$$

$$= \left[\frac{3000 + 0.626s}{s}\right]\left[\frac{9 \times 10^6 + 1879s + 1.142\,s^2}{s^2}\right]. \tag{3.55}$$

Inverting (3.52) gives the following transfer function, $H(s) = 1/A(s)$ as

$$H(s) = \frac{1}{A(s)} = \left[\frac{s}{0.626s + 3000}\right]\left[\frac{s^2}{1.142s^2 + 1879s + 9 \times 10^6}\right]. \tag{3.56}$$

Divide $H1(s)$ above and below by 0.626 and 1.142. (This works out as $1.3988 \cong 1.4$ on top.) This ensures the coefficient of the highest power of s in both factors is unity.

$$H(s) = \left[\frac{1.4s}{s + 4807}\right]\left[\frac{s^2}{s^2 + 1644s + 7.878 \times 10^6}\right]. \tag{3.57}$$

Place a **PORTBOTH-L** part and type in +Vcc for the positive supply and −Vcc for the negative supply. The names have no significance other than to tell PSpice to connect the two points together. From the Analysis Setup/AC Sweep set **Logarithmic, Point/Decade** = 1001, **Start Frequency** = 10, and **End Frequency** = 10k.

(2) A low-pass filter has the following specification:
- The maximum passband loss $A_{max} = 0.5$ dB
- The minimum stopband loss $A_{min} = 12$ dB
- The passband edge frequency $\omega_p = 100$ rs^{-1}
- The stopband edge frequency $\omega_s = 400$ rs^{-1}

Determine a Butterworth transfer function to meet this specification and simulate to verify the design.

CHAPTER 4

Infinite Gain Multiple Feedback Active Filters

4.1 MULTIPLE FEEDBACK ACTIVE FILTERS

The infinite gain multiple feedback (IGMF) active low-pass active filter circuit in Fig. 4.1 is required to meet a specification. This specification contains the passband and stopband gains A_{max}, A_{min}, and the passband and stopband edge frequencies ω_c and ω_s. There are two feedback paths hence the name.

After design and simulation, we see if the specification is met. The cut-off frequency $\omega_c = 1/C_1 R_1$ will enable you to set frequency range correctly in the **Analysis Setup**. From the **Analysis Setup**, select **AC Sweep** and then **Logarithmic, Point/Decade** = 1001, **S̲tart Frequency** = 1, and **E̲nd Frequency** = 10k. **Press F11** to simulate and check the filter specification from the frequency response in Fig. 4.2.

The impulse response may be obtained by applying an impulse to the input as in previous circuits.

4.2 SIMULATION FROM A NETLIST

An alternative to simulating from a schematic is to use a netlist with the file extension .net.

Press the little blue triangle. You might need to add the variables from the **Trace/Add Trace** menu if nothing appears on the Probe screen.

4.3 CHEBYCHEV IGMF BANDPASS FILTER

Fig. 4.5 shows an infinite-gain multiple-feedback filter configured as a bandpass filter.

The circuit attempts to use the full gain of the operational amplifier with feedback applied via C_1 and R_3. This circuit may be reconfigured as low-pass and high-pass active filters. The maximum Q-factor for this configuration is 25 but is generally designed for Q-factors in the 2–5 region. Let the capacitors have equal values, i.e., $C_1 = C_2 = C$ and obtain the transfer function from nodal analysis applied to node v (the junction of R_1, R_2 C_1 and C_2) as

$$v\left[\frac{1}{R_1} + \frac{1}{R_2} + 2s\,C\right] - \frac{E_i}{R_1} = s\,C E_0.$$

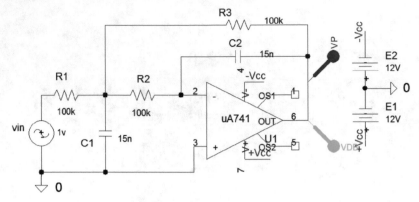

FIGURE 4.1: IGMF active second-order LPF

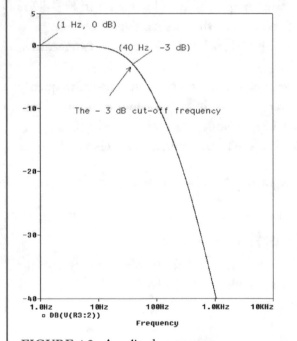
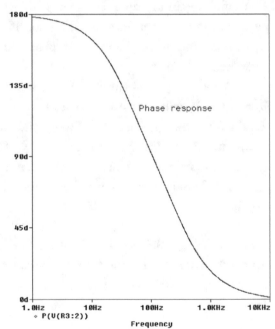

FIGURE 4.2: Amplitude response

But

$$\frac{E_0}{v} = \frac{-R_3}{1/sC} \Rightarrow v = -\frac{E_0}{sCR_3} \Rightarrow -\frac{E_0}{sCR_3}\left[\frac{1}{R_1} + \frac{1}{R_2} + 2sC\right] - \frac{E_i}{R_1} = sCE_0.$$

Divide across by E_0

$$-\frac{1}{sCR_3}\left[\frac{1}{R_1} + \frac{1}{R_2} + 2sC\right] - \frac{E_i}{E_0R_1} = sC \Rightarrow -\frac{R_1}{sCR_3}\left[\frac{1}{R_1} + \frac{1}{R_2} + 2sC\right] - \frac{E_i}{E_0} = sCR_1.$$

Open
Look in:
FIGURE4-001
AC
FIGURE4-001.net
My Network Places
File name:
FIGURE4-001.net
Files of type:
Probe Data Files (*.dat)
Open as:
Auto
Open
Cancel

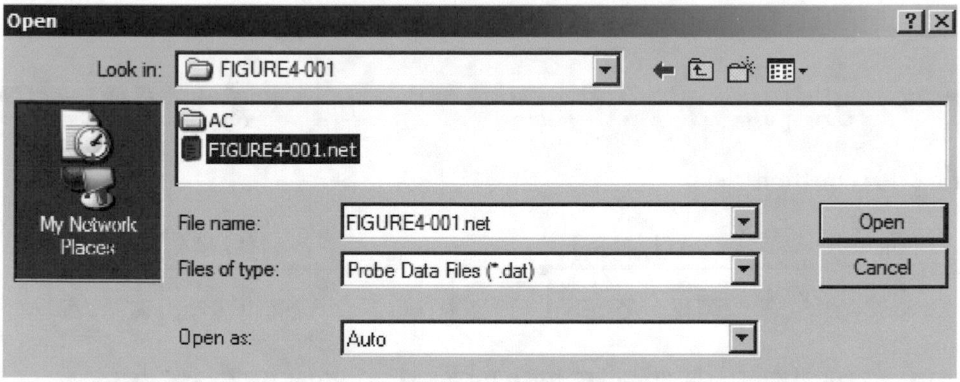

FIGURE 4.3: Select lpf.cir

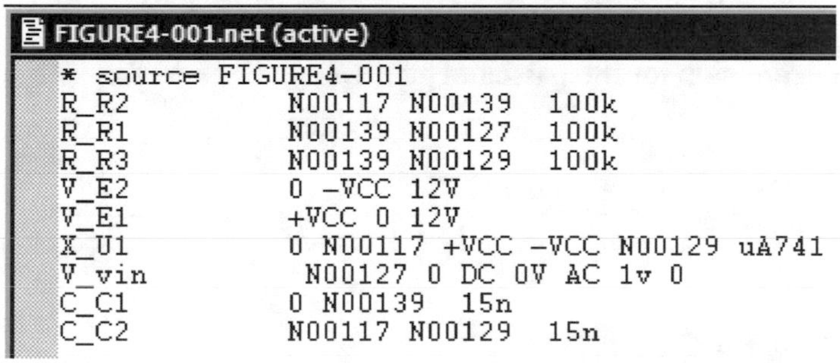

FIGURE4-001.net (active)
* source FIGURE4-001
R_R2 N00117 N00139 100k
R_R1 N00139 N00127 100k
R_R3 N00139 N00129 100k
V_E2 0 -VCC 12V
V_E1 +VCC 0 12V
X_U1 0 N00117 +VCC -VCC N00129 uA741
V_vin N00127 0 DC 0V AC 1v 0
C_C1 0 N00139 15n
C_C2 N00117 N00129 15n

FIGURE 4.4: The netlist

C1
15n
R3
107k
U1
-Vcc
R1
1k
v
OS1
C2
15n
Rsource
50
uA741
OUT
vout
VDB
R2
10Meg
OS2
vin
1v
+Vcc
0
+Vcc
E1
12V
12V
E2
-Vcc
0

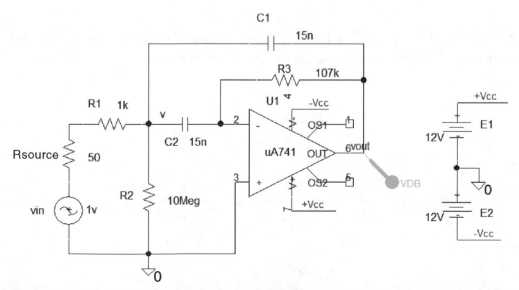

FIGURE 4.5: Infinite-gain multiple-feedback bandpass filter

So

$$\frac{E_i}{E_0} = -\frac{R_1}{sCR_3}\left[\frac{1}{R_1} + \frac{1}{R_2} + 2sC\right] - sCR_1 = -\frac{1}{sCR_3}\left[1 + \frac{R_1}{R_2} + 2sCR_1\right] - sCR_1.$$

Inverting this equation yields

$$\frac{E_0}{E_i} = -\frac{s(1/CR_1)}{s^2 + s\frac{R_1+R_1}{CR_1R_3} + \frac{1}{C^2R_1R_3}\left[1 + \frac{R_1}{R_2}\right]} = -\frac{s(1/CR_1)}{s^2 + s\frac{2}{CR_3} + \frac{1}{C^2R_3}\left[\frac{1}{R_1} + \frac{1}{R_2}\right]}.$$

Compare this equation to the standard form for a second-order bandpass, gives

$$\omega_o^2 = \frac{1}{C^2R_3}\left[\frac{1}{R_1} + \frac{1}{R_2}\right] \Rightarrow f_o^2 = \frac{1}{2\pi C^2 R_3}\left[\frac{1}{R_1} + \frac{1}{R_2}\right].$$

To determine an expression for the passband gain $G(\omega_0)$, but $s = j\omega$, so the transfer function is

$$\frac{E_0}{E_1} = -\frac{j\omega_0 1/CR_1}{-\omega_0^2 + \omega_0\frac{1}{R_3}\left[\frac{2}{C}\right] + \omega_0^2} = G(\omega_0) = -\frac{R_3}{2R_1}.$$

4.3.1 Expressions for Q and BW

$$\frac{\omega_0}{Q} = \frac{2}{CR_3} \Rightarrow Q = \frac{\omega_0 C R_3}{2} \Rightarrow R_3 = \frac{2Q}{\omega_0 C}.$$

The -3 dB bandwidth is $BW = 1/\pi CR_3$ and we need to determine values for the resistors using the derived expressions. This configuration places certain limitations on the value of R_2. Let $G(\omega_0) = G$ then

$$R_1 = -\frac{R_3}{2G} = \frac{Q}{G\omega_0 C}.$$

Thus

$$\omega_0^2 = \frac{4Q^2}{R_3} \Rightarrow Q = \frac{\omega_0 C R_3}{2} = \frac{1}{C^2 R_3}\left[\frac{1}{R_1} + \frac{1}{R_2}\right].$$

Therefore

$$\frac{4Q^2}{R_3} = \frac{1}{R_1} + \frac{1}{R_2}.$$

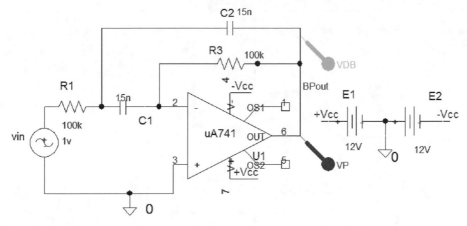

FIGURE 4.6: IGMF Bandpass filter without R2

Substituting for R_1

$$\frac{4Q^2}{R_3} = \frac{G\omega_0 C}{Q} + \frac{1}{R_2}.$$

Substituting for R_3

$$\frac{2Q^2\omega_0 C - G\omega_0 C}{Q} = \frac{1}{R_2} \Rightarrow \frac{Q}{2Q^2\omega_0 C - G\omega_0 C} = R_2.$$

To avoid negative values for R_2 we must ensure $2Q > G$. High gain means high selectivity so to overcome this, remove R_2 and let $R_1 = 100$ kΩ, $R_3 = 100$ kΩ, and $C_1 = C_2 = 15$ nF, as shown in Fig. 4.6.

From the **Analysis Setup,** select **AC Sweep** and **Linear, Point/Decade** =1001, **Start Frequency** = 100, and **End Frequency** = 10k. Press **F11** to simulate and use the cursors to measure the −3 dB bandwidth as shown from the frequency response in Fig. 4.7. Verify all calculations, i.e., passband gain, Q-factor etc.

The transfer function then becomes

$$\frac{E_0}{E_i} = -\frac{s/R_1 C}{s^2 + s(2/C R_3) + 1/(C^2 R_1 R_3)}. \tag{4.1}$$

Here the resonant frequency is $\omega_0^2 = 1/(C^2 R_1 R_3)$ the Q-factor is $Q = 0.5\sqrt{R_3/R_1}$, and the passband gain is $G(\omega_0) = R_3/2R_1 = 2Q^2$.

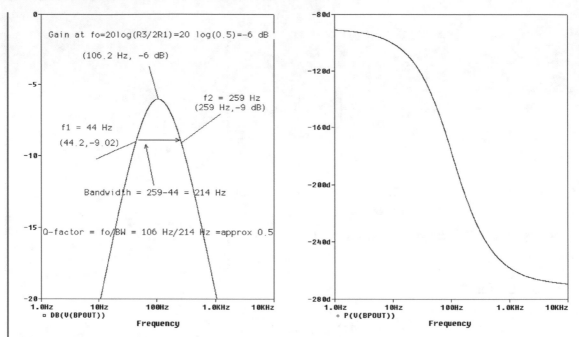

FIGURE 4.7: The IGMF frequency response for R2 = ∞

4.4   CHEBYCHEV BANDPASS ACTIVE FILTER

Obtain a Chebychev bandpass transfer function to meet the following filter specification:

- The maximum passband attenuation $A_{\mathrm{max}} = 0.5$ dB
- The minimum stopband attenuation $A_{\mathrm{min}} = 5$ dB
- The lower stopband edge frequency $f_{s1} = 1000$ Hz
- The upper stopband edge frequency $f_{s2} = 3000$ Hz
- The lower passband edge frequency $f_{p1} = 1800$ Hz
- The upper passband edge frequency $f_{p2} = 2200$ Hz

The setup in Fig. 4.8 is used to compare the theoretical transfer function to the circuit implementation.

Obtain suitable component values for C and R_2 to meet the specification, if $R_1 = 1$ kΩ. The Chebychev filter order is

$$
n = \frac{\cosh^{-1}\left[\frac{10^{0.1\,A_{min}}-1}{10^{0.1\,A_{max}}-1}\right]^{1/2}}{\cosh^{-1}(\Omega_S)}
$$

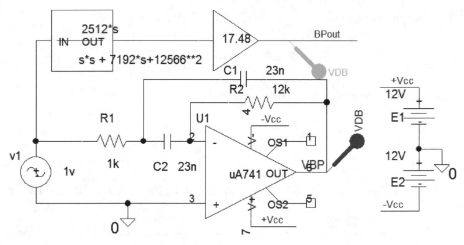

FIGURE 4.8: IGMF bandpass active filter

where $\Omega_s = -[\frac{f_{s2}-f_{s1}}{f_{p2}-f_{p1}}] = [\frac{3000-1000}{2200-1800}] = \frac{2000}{400} = 5$, hence, $n = \frac{\cosh^{-1}\left[\frac{10^{0.1\times15}-1}{10^{0.1\times0.5}-1}\right]^{1/2}}{\cosh^{-1}(5)} = 0.92 \approx 1$. Select a first-order approximation loss function $A(\$) = \$ + \mathbf{2.863}$ from the Chebychev tables where the ripple is 0.5 dB maximum ripple in the passband. $\$$ is the normalized complex frequency variable and the resonant frequency is calculated:

$$f_0 = \sqrt{f_{p1}f_{p2}} = \sqrt{1800*2200} = 2000 \text{ Hz}, \quad \text{or} \quad 2\pi 2000 = 12566 \text{ rs}^{-1}.$$

Apply the bandpass frequency transformation $\$ = (s^2 + \omega_0^2)/Bs$ to the loss function. (Note: inverting this transform equation will lead to a bandstop filter.) B is the −3 dB bandwidth expressed as

$$B = \omega_{p2} - \omega_{p1} = 2\pi(400) = 2512 \text{ rs}^{-1}.$$

Substitute these values into the bandpass frequency transform equation $\$ = (s^2 + \omega_0^2)/Bs = (s^2 + 12566^2)/2512s$ and then denormalize the loss function as $A(s) = [(s^2 + 12566^2)/2512]s + 2.863$. The transfer function is obtained by inverting the denormalized loss function as

$$H(s) = \frac{1}{A(s)} = \frac{2512s}{s^2 + 7192s + 12566^2} = \frac{ks}{s^2 + \omega_0/Qs + \omega_0^2}$$

$$= -\frac{s1/R_1C}{s^2 + s(2/CR_3) + 1/(C^2R_1R_3)}.$$

We may plot this transfer function using a **Laplace** part. What is now left is to get values for the component and we do this by comparing the coefficients of this function to the coefficients from (5.81) that yields ω_0/Q, and $\omega_0^2 = 1/C^2 R_1 R_3$. Let $R_1 = 1$ kΩ.

$$R_3 = \frac{1}{C^2 10^3 \times 12566^2}.$$

The capacitance is determined from the s coefficient as

$$7192 = \frac{2}{C R_3} \Rightarrow C = \frac{2}{R_3 7192}$$

$$R_3 = \frac{1}{\frac{2}{R_3 7192} 10^3 \times 12566^2} = \frac{1}{\frac{4}{(R_3 7192)^2} 10^3 \times 12566^2} = \frac{(R_3 7192)^2}{4 \times 10^3 \times 12566^2}$$

$$1 = \frac{R_3 (7192)^2}{4 \times 10^3 \times 12566^2} \Rightarrow R_3 = \frac{4 \times 10^3 \times 12566^2}{(7192)^2} = 12 \text{ k}\Omega.$$

Substituting for R_3 back into the previous equation gives a value for the capacitance as

$$C = \frac{2}{12210 \times 7192} \approx 23 \text{ nF}.$$

From the **Analysis Setup,** select **AC Sweep/Logarithmic**, **Point/Decade** $= 1001$, **S̲tart Fre-quency** $= 10$, and **E̲nd Frequency** $= 10$k. Press **F11** to simulate and check, using the cursors as in Fig. 4.9, that the specification is met.

The gain is $G(\omega_0) = R_3/2 R_1 = 12k/2k = 6$, or in dB $= 20 \log(6) = 15.5$ dB. A **GAIN** part is inserted after the **Laplace** part in order to make the gain equal to $2512/(1/R_1 C) = 17.48$ so that both transfer functions have the same passband gain. The bandwidth is not -3 dB in this case but where the output falls by 0.5 dB from the specification. Also, the attenuation at the stopband edge frequency is measured as 6.7 dB- an attenuation of 15.5 dB- 6.7 dB $= 8.8$ dB which is better than the specification demands.

4.5   EXAMPLE 2

Fig. 4.10 is a third-order active filter formed by cascading first- and second-order active filters.

The passband edge frequency is $\omega_p = 1000$ rs^{-1} and the frequency correction factor is $\varepsilon = 1$. A normalized third-order Butterworth loss function $A(\$) = (\$ + 1)(\$^2 + \$ + 1)$ is de-normalized by substituting $\$ = s/\omega_p = s/1000$. Invert the de-normalized loss function to give the required transfer function

$$H(s) = \frac{1}{\left(\frac{s}{1000} + 1\right)\left(\frac{s^2}{10^6} + \frac{s}{10^3} + 1\right)} = \frac{(1000)(10^6)}{(s + 1000)(s^2 + 1000s + 10^6)}. \qquad (4.2)$$

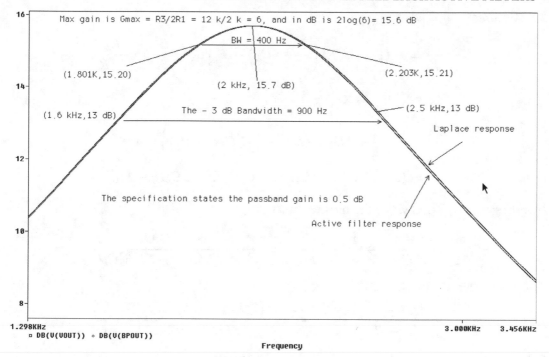

FIGURE 4.9: Infinite-gain multiple-feedback active filter bandpass response

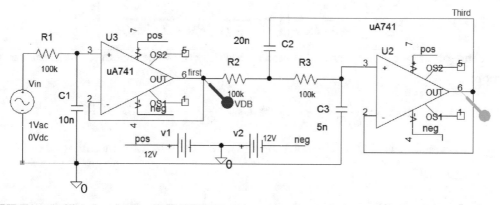

FIGURE 4.10: Third-order S and K LPF

Apply nodal analysis to the node junction of R_2 and R_3, but let the resistors have a value R.

$$v\left(\frac{1}{R} + \frac{1}{R} + s\,C_2\right) - E_3 s\,C_2 - \frac{E_3}{R} = \frac{E_2}{R}.$$

From the potential divider rule, we may write the relationship between v and E_3 as

$$v\left(\frac{1/s\,C_3}{1/s\,C_3 + R}\right) = E_3 \Rightarrow v = E_3(1 + s\,C_3 R).$$

Substituting for v yields

$$\frac{E_2}{R} = E_3(1 + s\,C_3\,R)\left(\frac{2}{R} + s\,C_2\right) - E_3 s\,C_2 - \frac{E_3}{R} \Rightarrow \frac{E_2}{E_3\,R}$$

$$= (1 + s\,C_3\,R)\left(\frac{2}{R} + s\,C_2\right) - s\,C_2 - \frac{1}{R}$$

$$\frac{E_2}{E_3} = R\left(\frac{2}{R} + s\,C_2 + s\,2C_3 + s^2 C_2 C_3 R - s\,C_2\right) - 1$$

$$= (2 + s\,C_2\,R + s\,2C_3\,R + s^2 C_2 C_3 R^2 - s\,C_2\,R) - 1.$$

Inverting this equation yields the transfer function

$$\frac{E_3}{E_2} = \frac{1}{s^2 C_2 C_3 R^2 + s\,2C_3\,R + 1} = \frac{1/R^2 C_2 C_3}{s^2 + s\,2/(C_2 R) + 1/(R^2 C_2 C_3)}.$$

The first-order transfer function is

$$\frac{E_2}{E_1} = \frac{1/s\,C_1}{1/s\,C_1 + R} = \frac{1}{1 + s\,C_1\,R} = \frac{1/RC_1}{s + 1/RC_1} = \frac{1000}{s + 1000} = \frac{\omega_p}{s + \omega_p}.$$

Substituting for R and ω_p yields $C_1 = \frac{1}{10^3 10^5} = 0.01\ \mu\text{F}.$

$$\frac{E_3}{E_1} = \left[\frac{1/RC_1}{s + 1/RC_1}\right]\left[\frac{1/R^2 C_2 C_3}{s^2 + (2/C_2 R)s + 1/R^2 C_2 C_3}\right].$$

The second-order components are calculated by equating the s coefficient terms in each transfer function:

$$1000 = \frac{2}{C_2 R} = \frac{2}{C_2 10^5} \Rightarrow C_2 = \frac{1}{500 \times 10^5} = 0.02\ \mu\text{F}.$$

The second capacitor is obtained by equating the non-s term in the denominators of (4.2) and (4.3) as

$$\omega_p^2 = \frac{1}{C_2 C_3 R^2} \Rightarrow C_3 = \frac{1}{C_2 R^2 \omega_p^2} = \frac{1}{0.02 \times 10^{-6} 10^{10} 10^6} = \frac{1}{0.2 \times 10^9} = 5\ \text{nF}.$$

4.5.1   Low-Pass Frequency Response

Place markers on the outputs of the first- and second-order stages. From the **Analysis Setup**, select **AC Sweep/Logarithmic, Point/Decade** = 1001, **Start Frequency** = 10, and **End Frequency** = 10k. Press **F11** to simulate. Does the filter response in Fig. 4.11 meet the specification?

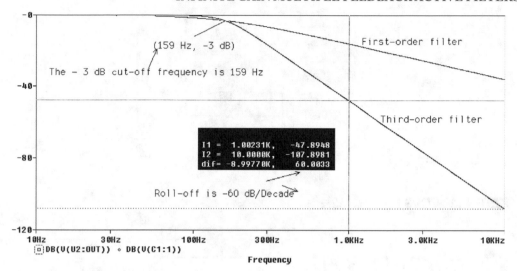

FIGURE 4.11: Frequency response of LPF

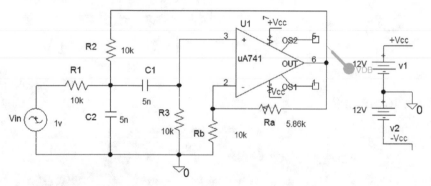

FIGURE 4.12: Bandpass VCVS filter

4.6   EXERCISES

(1) Obtain the transfer function for the bandpass Sallen and Key filter circuit in Fig. 4.12 and work out an expression for the passband edge frequency.

To minimize the degrees of freedom, let $C_1 = C_2 = C$ and $R_1 = R_2 = R$. Measure the bandwidth from the bandpass filter response shown in Fig. 4.13, hence calculate the Q factor and simulate.

(2) Obtain the transfer function and an expression for the passband edge frequency for the circuit in Fig. 4.13 and investigate its performance by simulating.

(3) Investigate the low-pass IGMF active filter in Fig. 4.14.

Obtain the transfer function assuming equal component values. Hence, obtain expressions for the cut-off frequency and passband gain. Investigate the advantages or

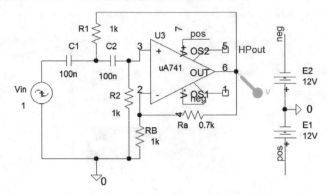

FIGURE 4.13: VCVS high-pass filter

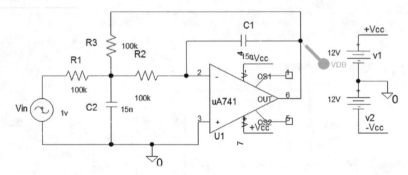

FIGURE 4.14: IGMF LPF

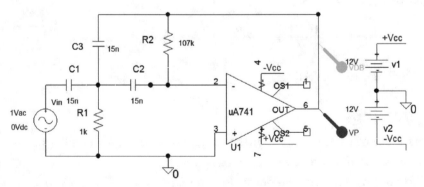

FIGURE 4.15: IGMF high-pass filter

disadvantages, between IGMF and VCVS filter types, by plotting the input and output impedances for both. Plot the feedback current and comment on the results.

(4) Investigate the high-pass IGMF active filter in Fig. 4.15.

(5) For the S and K active filter, sweep Ra and observe how the filter turns into an oscillator for $k = 3$.

CHAPTER 5

Biquadratic Filters and Monte-Carlo Analysis

5.1 BIQUAD ACTIVE FILTER

The biquad circuit in Fig. 5.1 has bandpass and low-pass filter outputs, with achievable very high Q-factors unlike the VSVS and the IGMF circuits which have small Q-factors. The biquad circuit, whilst not as useful as the state variable active filter, nevertheless has certain applications. It is easily tunable using a stereo potentiometer (ganged). The damping factor (damping factor is equal to the inverse of the Q-factor), configures the filter for a Butterworth or Chebychev responses. We have used an ideal opamp in the last stage to satisfy the demo version requirements.

The resistances in the last two stages are made variable and ganged together. In reality, two resistances are ganged by manufacturing them on the same spindle. To simulate "ganging" in PSpice, we use a **PARAM** part making the second variable resistor dependent on the first variable resistance {**Rvar1**}. This is necessary since we may sweep only one component at any time in PSpice.

5.1.1 The Biquad Bandpass Transfer Function

The biquad active filter consists of a leaky integrator, an integrator, and a summing amplifier. To simplify the analysis, let $R_1 = R_2 = R_3 = R = 10$ kΩ, and $C_1 = C_2 = C = 15$ nF, with the parallel combination of R and C equal to $Z_f = \frac{R}{1+sCR}$. The input is V_1 and the last output is V_4. We derive the bandpass transfer function V_2/V_1 first. The first op-amp is a leaky integrator, with the output voltage defined as

$$V_2 = -\frac{Z_f}{R}V_4 - \frac{Z_f}{R}V_1. \tag{5.1}$$

We must eliminate V_4 in (5.1) so the transfer function for the last stage is

$$\frac{V_4}{V_3} = -\frac{R_7}{R_6} = -\frac{R}{R} = -1 \Rightarrow V_4 = -V_3. \tag{5.2}$$

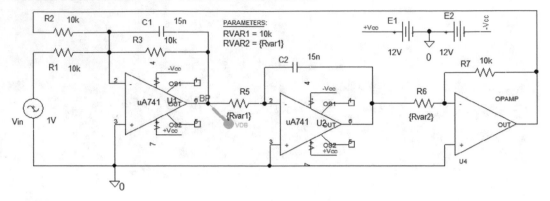

FIGURE 5.1: Biquad filter

The gain for the second stage is

$$\frac{V_3}{V_2} = -\frac{1}{sCR}.$$ (5.3)

From (5.2) and (5.3) we write

$$V_4 = \frac{1}{sCR}V_2.$$ (5.4)

Substituting (5.4) into (5.1) yields the transfer function for the bandpass output as

$$V_2 = -\frac{Z_f}{R}\frac{1}{sCR}V_2 - \frac{Z_f}{R}V_1 \Rightarrow V_2\left[1 + \frac{Z_f}{sCR^2}\right] = -\frac{Z_f}{R}V_1.$$ (5.5)

Substituting Z_f into (5.5) yields

$$\frac{V_2}{V_1} = -\frac{Z_f/R}{1 + Z_f/sCR^2} = -\frac{R/(1+sCR)}{R + R/(1+sCR)/sCR} = -\frac{s/CR}{s^2 + s/CR + 1/C^2R^2}.$$ (5.6)

The standard bandpass second-order function is

$$\frac{V_2}{V_1} = -\frac{sK(\omega_p/Q)}{s^2 + s(\omega_p/Q) + \omega_p^2}.$$ (5.7)

Expressions for the passband edge frequency, Q-factor, and gain are obtained in terms of circuit components by comparing denominator coefficients of (5.6) and (5.7). The resonant frequency is

$$\omega_o = \frac{1}{CR} = \frac{1}{10 \times 10^{-9}10^4} = 6666 \text{ rs}^{-1}.$$ (5.8)

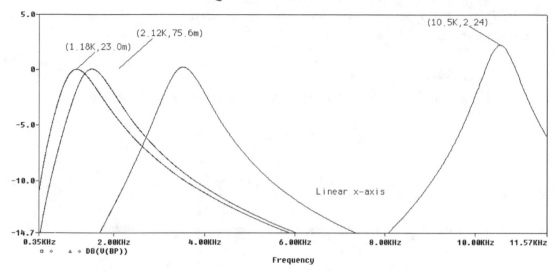

FIGURE 5.2: Biquad filter response

An expression for the gain H at resonance is obtained by replacing s with $j\omega_0$.

$$H(\omega_0) = \left.\frac{s\,6666}{s^2 + s\,6666 + 6666^2}\right|_{s=j\omega_o} = \frac{j\omega_o\,6666}{(j\omega_o)^2 + j\omega_o\,6666 + 6666^2}$$

$$= \frac{j\omega_o\,6666}{-\omega_o^2 + j\omega_o\,6666 + 6666^2} = \frac{6666}{6666} = 1. \qquad (5.9)$$

The gain in dB at resonance is $20\log(1) = 0$ dB, and the Q-factor is

$$Q = \sqrt{\frac{C_1 R_2^2}{C_2 R_6 R_3}}. \qquad (5.10)$$

The -3 dB bandwidth BW is

$$BW = \frac{1}{2\pi C_1 R_2}. \qquad (5.11)$$

Fig. 5.2 shows how the Q-factor and resonant frequency are not independent of each other so that for high frequencies, the bandwidth is the same as that for low frequencies. This, in general, is not a desirable feature. For example, equalizing a signal in an audio mixing desk is normally implemented with a state-variable circuit, where the bandwidth changes with the higher frequencies. The tuning procedure for the biquad BP is to select values for C_1, C_2 and R_5, and adjust the resonant frequency ω_p, by varying R_3, with the gain set by R_1. The Q-factor is set by R_2 and not by an iterative approach where you keep adjusting the component to get the desired response.

5.1.2   Equal Value Component

Let $C_1 = C_2 = C$ and $R_3 = R_6 = R$ (leads to a poor unity Q-factor, however)

$$\frac{V_2}{V_1} = -\frac{s/CR}{s^2 + s/CR + 1/C^2R^2} \tag{5.12}$$

$$H(0) = K = R_2/R_1 = 1 \tag{5.13}$$

$$\omega_o = \frac{1}{CR} \Rightarrow f_o = \frac{1}{2\pi CR} = \frac{1}{2\pi 15 \times 10^{-9}10^4} = 1136 \text{ Hz} \tag{5.14}$$

$$Q = \frac{R}{R} = 1 \tag{5.15}$$

$$BW = 1/2\pi CR = 1136 \text{ Hz}. \tag{5.16}$$

From Analysis Setup/AC Sweep menu, set **Logarithmic, Point/Decade** = 1001, **Start Frequency** = 10, and **End Frequency** = 10k. From the Analysis menu, set the parametric parameters: **Swept Var Type** = Global parameters, **Sweep Type** = **Linear**, **Name** = Rvar1, **StartValue** = 1k, **End Value** = 10k, **Increment** = 2k. Press **F11** to simulate, and, using the cursors, measure the filter parameters from the frequency response shown in Fig. 5.2. From the measurements calculate the Q-factor using f_0/BW.

You may change the Probe output display by selecting **Tools/Options**. In the **Use Symbols** section, experiment with the various options available. For example, selecting **Always** produces a cluttered display for a swept component but is useful nevertheless, for certain applications.

5.2   STATE-VARIABLE FILTER

The state-variable filter in Fig. 5.3 is similar to the biquad filter but is more useful in that it achieves high Q-factors, but also has a high pass output.

A popular application for this circuit is the equalizer section of an audio mixing desk. The transfer functions for the three stages are: High-pass = $-V_2/V_1$, bandpass = $-V_3/V_1$, and low-pass = $-V_4/V_1$. The resistance of each integrator section is variable and ganged on the same spindle. PSpice 'ganging' uses a **PARAM** part to make the second resistance dependent on the first as **Rvar1** = {**Rvar1**}. The first high-pass output V_2 is obtained by applying the principle of superposition to the input voltages applied to R_1 and R_2 and the noninverting

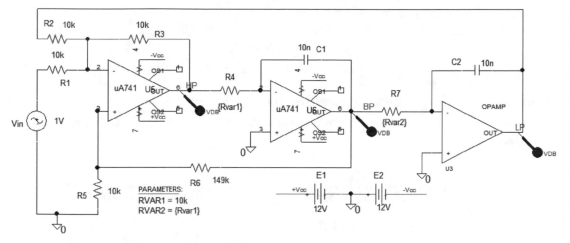

FIGURE 5.3: State-variable filter

input. Let $R_1 = R_2 = R_3 = R = 10$ kΩ.

$$V_2 = -\frac{R_2}{R_1}V_1 - \frac{R_2}{R_1}V_4 + V_3 \left\{ \frac{R_5}{R_5 + R_6} \right\} \left\{ 1 + \frac{R_3}{R_1/2} \right\}.$$

Let $R_4 = R_7 = R = 10$ kΩ and $C_1 = C_2 = C = 10$ n. The bandpass output voltage is $V_3 = -V_2/sCR$, and the LP output is $V_4 = -V_3/sCR$. Hence, $V_4 = (-\frac{1}{sCR})(-\frac{1}{sCR})V_2 = \frac{V_2}{s^2 C^2 R^2}$. Substitute for V_4 into the first equation to yield

$$V_2 = -\frac{R_2}{R_1}V_1 - \frac{R_2}{R_1}\frac{V_2}{s^2 C^2 R^2} - \frac{V_2}{sCR}\left(\frac{R_5}{R_5 + R_6}\right)(1+2)$$

$$\frac{V_2}{V_1} = -\frac{s^2}{s^2 + s\frac{3}{CR}\left(\frac{R_5}{R_5 + R_6}\right) + \frac{1}{C^2 R^2}}.$$

The bandpass output (BP) transfer function is obtained by substituting for V_4, V_3, $R_5 = 10$ kΩ, and $R_6 = 149$ kΩ.

$$\frac{V_3}{V_1} = \frac{s1/CR}{s^2 + s\frac{3}{CR}\left(\frac{R_5}{R_5 + R_6}\right) + \frac{1}{C^2 R^2}} = \frac{s10^4}{s^2 + s3 \times 10^4(0.0629) + 10^8} = \frac{s10^4}{s^2 + s1886 + 10^8}.$$

Similarly for the low-pass transfer function

$$\frac{V_4}{V_1} = -\frac{1/C^2 R^2}{s^2 + s\frac{3}{CR}\left(\frac{R_5}{R_5 + R_6}\right) + 1/C^2 R^2}.$$

The standard form for a second-order bandpass function is

$$\frac{V_{out}}{V_{in}} = -\frac{s\,K\omega_p/Q}{s^2 + s\omega_p/Q + \omega_p^2}.$$

Write expressions for the passband edge frequency, Q-factor and gain, in terms of the circuit components, by comparing the last two transfer functions denominator coefficients for the low-pass output:

$$H(\omega_0) = \frac{s\,10^4}{s^2 + s\,1886 + 10^8}\bigg|_{s=j\omega_0} = \frac{j\omega_0 10^4}{(j\omega_0)^2 + j\omega_0 1886 + 10^8}$$

$$= \frac{j\omega_0 10^4}{-\omega_0^2 + j\omega_0 1886 + 10^8} = \frac{10^4}{1886} = 5.3.$$

The gain in dB at resonance is $20\log(5.3) = 14.48$ dB.

$$\omega_0 = \frac{1}{CR} \Rightarrow f_0 = \frac{1}{2\pi CR} = \frac{1}{2\pi 10 \times 10^{-9}10^4} = 1591 \text{ Hz}.$$

The Q-factor is

$$Q = \frac{1}{3}\frac{R_5 + R_6}{R_5} = 0.333\frac{159k}{10k} = 5.29.$$

The bandwidth is

$$BW = \frac{f_0}{Q} = \frac{1591}{5.29} = \frac{3}{2\pi CR}\left(\frac{R_5}{R_5 + R_6}\right) = \frac{3}{2\pi 10^{-9}10^4}\left(\frac{10k}{159k}\right) = 300 \text{ Hz}.$$

The Q-factor and resonant frequency are independent of each other (unlike the biquad circuit). From the **Analysis Setup,** select **AC Sweep, Linear, Point/Decade** = 1001, **Start Frequency** = 300 and **End Frequency** = 30k. Press **F11** to simulate and measure the bandwidth from the bandpass filter response in Fig. 5.4. The **Evaluate measurement or measurement** functions from the **Trace** menu in Probe contains useful automatic plotting routines, such as **Bandwidth_Bandpass_3dB(V(BP))** for the bandwidth where the results appear at the bottom of the trace.

5.2.1   Bandpass Frequency Response

We may repeat the last simulation using the **Parametric** sweep from the Analysis setup menu to vary the value of Rvar to produce the three bandpass frequency responses as shown in Fig. 5.5.

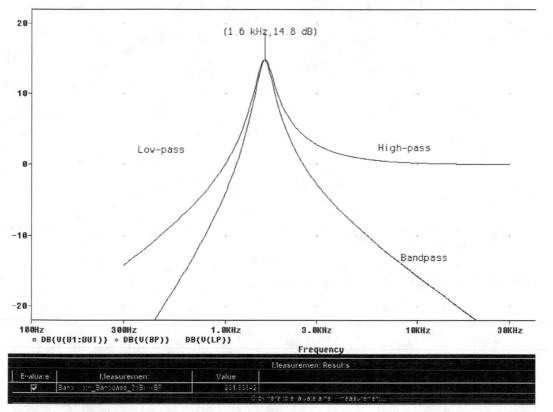

FIGURE 5.4: State-variable filter frequency response

5.3   MONTE CARLO ANALYSIS AND HISTOGRAMS

In the real electronics world, resistors, capacitors, operational amplifiers etc. have parameter values that differ from their ideal values. Thus, a 20 % 1 kΩ resistor tolerance could have a value that deviates by 200 Ω. Sensitivity analysis allows us to predict circuit behavior when using these resistors. This analysis is tedious but, thankfully, PSpice comes to the rescue. Higher-order filters should be designed using components with 1 % or less, tolerances. In the following exercise, we place a tolerance on all the resistors in a sixth-order active low-pass filter and use **Monte-Carlo** (or **Worst-case**) facility in Probe to show how the component tolerance can produce a range of frequency responses. The results can then be displayed in **Histogram** form.

5.3.1   Sixth-Order LPF Active Filter

Performance Analysis applies a Monte-Carlo performance analysis to the sixth-order active LPF filter in Fig. 5.6 to investigate the effect of resistor component tolerance on the frequency

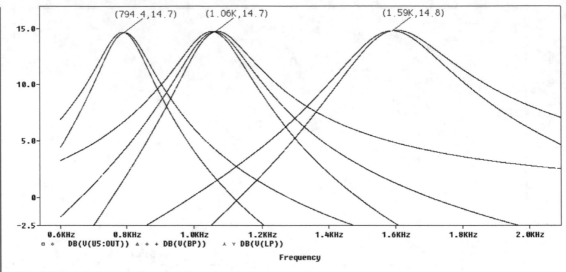

FIGURE 5.5: Making R4 and R7 variable and ganged

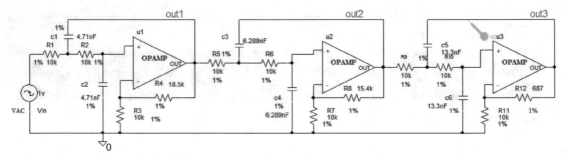

FIGURE 5.6: Sallen and Key active second-order LPF

response. Ideal amplifier parts called **opamp** are used in this design to simplify the process. The active filter specification is:

- The maximum passband loss $A_{max} = 1$ dB
- The minimum stopband loss $A_{min} = 15$ dB
- The passband edge frequency $f_p = 3.4$ kHz
- The stopband edge frequency $f_s = 4$ kHz

A Chebychev filter design is considered and we calculate the filter order as

$$n = \frac{\cosh^{-1}[10^{0.1\,A_{min}} - 1)/(10^{0.1\,A_{max}} - 1)]^{0.5}}{\cosh^{-1}(\Omega_s)} = \frac{\cosh^{-1}[10^{0.1\times15} - 1)/(10^{0.1\times1} - 1)]^{0.5}}{\cosh^{-1}(4/3.4)} \approx 6.$$

The passband gain for each stage is $G(0) = 20 \log(1 + R_a/R_b)$ and component values for the three stages are calculated using the process outlined in previous chapters and are:

- First stage: $R_1 = R_2 = 10 \text{ k}\Omega$, $C_1 = C_2 = 4.71 \text{ nF}$, $R_b = 10 \text{ k}\Omega$, $Ra = 18.5 \text{ k}\Omega$
- Second stage: $R_3 = R_4 = 10 \text{ k}\Omega$, $C_3 = C_4 = 6.28 \text{ nF}$, $R_d = 10 \text{ k}\Omega$, $R_c = 15.4 \text{ k}\Omega$
- Third stage: $R_5 = R_6 = 10 \text{ k}\Omega$, $C_5 = C_6 = 13.3 \text{ nF}$, $Rf = 10 \text{ k}\Omega$, $R_e = 687 \text{ }\Omega$

5.3.2   Component Tolerance

Sixth-order active filters should be designed using one percent tolerance resistors (typical lab resistors are 20%). The capacitors and op-amps are also nonideal, but we will limit this Monte-Carlo analysis to the resistors only. The components are given a tolerance value for several analysis "**runs**" on an AC frequency response. Each run varies the model parameter randomly, but within the tolerance range (hence the name Monte-Carlo as in gambling). The first run uses the face value of the component. In this circuit we need to give a tolerance to all resistors. To change the resistor tolerance, select a resistor, hold down the **ctrl** key, and select each resistor in turn. **Rclick** and select **Edit Properties**, and in the **Tolerance** row enter 20% for each resistor (you must include the percentage symbol) as in Fig. 5.7.

Repeat this procedure for the capacitors. Select the **Edit Simulation** Settings. To run a Monte Carlo analysis, ticking **Monte Carlo/Worst Case** as in Fig. 5.8 displays the MC parameters.

The Monte-Carlo parameters are: **MC Runs** = enter the number of runs = 75, **Analysis Type** = **AC**, **Output Variable** is the name of the output node **out3** entered as **V(out3)**. The **More Settings** menu produces a further menu also shown in Fig. 5.8 where you can select from a range of tasks, but here we select the greatest difference in the y-axis variable. After

TOLERANCE	20%	20%	20%	20%

FIGURE 5.7: Setting the resistor tolerance

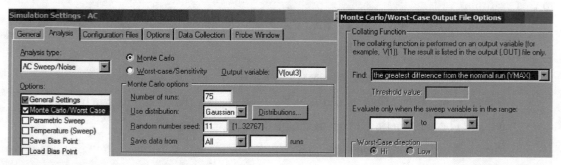

FIGURE 5.8: Monte Carlo parameters

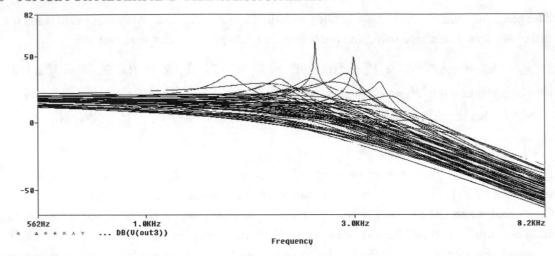

FIGURE 5.9: Note the variation on the output voltage out3

simulation, 75 runs are plotted, so select all. The last stage output response is plotted for 75 runs in Fig. 5.9.

5.4   PERFORMANCE HISTOGRAMS

Performance histograms are bar charts showing the number of occurrences of an event for each of several divisions of the *y* variable. The *x*-axis becomes divisions of the *y* variable, and the *y*-axis becomes the number of occurrences. In **Performance Analysis**, the measurement function is calculated from each Monte Carlo run and is an occurrence. Only one histogram is allowed per plot. Performance analysis plots a histogram of the −3 dB point to show how this parameter changes with tolerance. Delete any output traces and from the **Trace/Performance Analysis menu,** select **Wizard** as shown in Fig. 5.10.

Select **Cutoff_Lowpass_3dB** as shown in Fig. 5.11.

Press the red trace icon shown in Fig. 5.12 and search for the output variable (V(out3) in this example).

Selecting **Finish** should display the Histogram and statistics as shown in Fig. 5.14.

The statistics displayed are

- *n* samples: The number of Monte Carlo runs.
- *n* divisions: The number of vertical bars in the *x*-axis that make up the histogram.
- Mean: The arithmetic average of the goal function values.
- Sigma: The standard deviation.
- Minimum: The minimum goal function value.

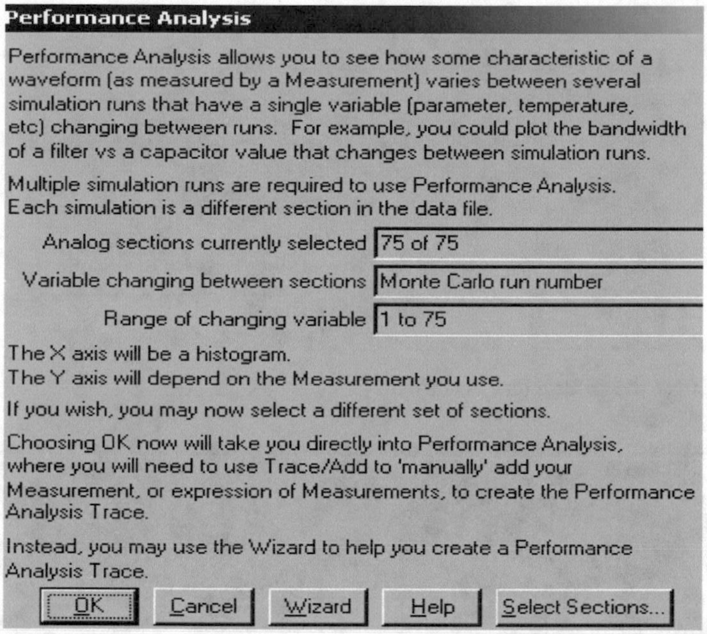

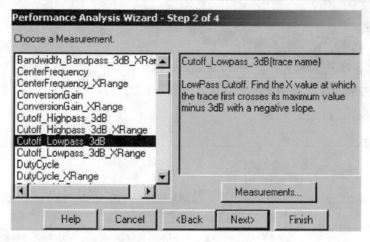

FIGURE 5.10: Select Wizard

FIGURE 5.11: Chose the measurement Function

- 10th %tile: 10% of the goal function values that are less than or equal to the number.

- Median: The same as the 10th percentile but the percentage is 50%.

- 90th %tile: The same as 10th percentile but the percentage is 90%.

- Maximum: The maximum goal function value.

Repeat this analysis but set the tolerance to 1% and observe the improvement in the response.

Performance Analysis Wizard - Step 3 of 4

Measurement Expression

Cutoff_Lowpass_3dB(V(out3))

Now you need to fill in the Measurement arguments. That is, you need to tell the Measurement which trace(s) to look at, and if necessary, the other numbers the Measurement needs to work.

The Measurement 'Cutoff_Lowpass_3dB' has 1 argument. Please fill it in now.

Name of trace to search   [⌇]   V(out3)

FIGURE 5.12: Enter the parameters as shown

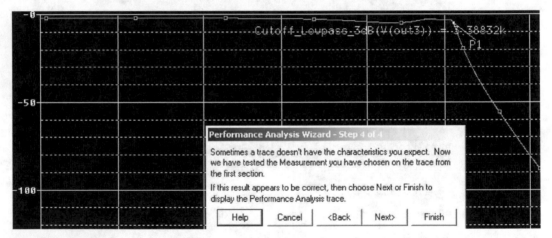

FIGURE 5.13: Select Finish

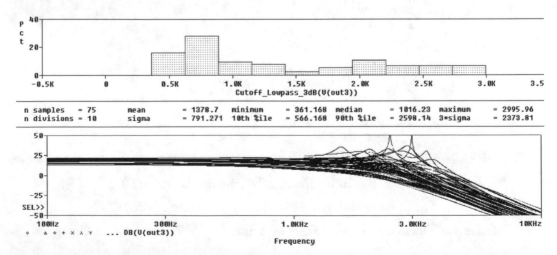

FIGURE 5.14: Gaussian distribution of the −3 dB point

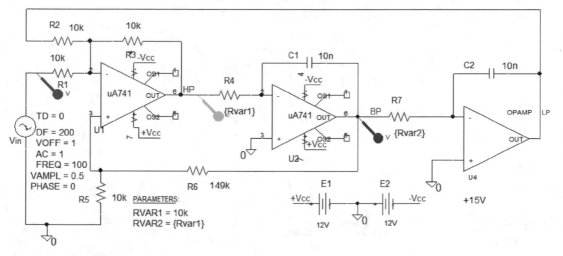

FIGURE 5.15: State-variable filter

5.5 EXERCISES

(1) In the state-variable filter configuration, add a further opamp configured as a two-input summing amplifier and connect the HP and LP outputs to show it produces a notch filter (a bandstop filter).

(2) Obtain the transfer function for the state-variable bandpass filter circuit. Assuming a value $R = 10$ kΩ, calculate a value for C, which sets the center frequency to 10 krs^{-1}. Calculate values for R_3 and R_4 to produce a Q-factor $= 50$. A fourth-order bandpass filter has a transfer function defined as

$$H(\$) = -\frac{\$^2}{(\$^2 + 0.765\$ + 1)(\$^2 + \$1.848\$ + 1)}.$$

Calculate suitable component values using two cascaded state-variable bandpass transfer functions.

(3) Repeat the above exercise, but attach a tolerance to each capacitor. The model is defined as: MODEL CTOL CAP (C=1 DEV/gauss 10%).

(4) This is just a bit of fun (and art?). Apply different signal types to the state variable in Fig. 5.15.
Carry out a transient analysis, and in Probe with the bandpass output selected, change the x-axis to the low-pass output, and then try the HP output. There are no rules so investigate different signal types and frequencies. Change the Q-factor of the filter and

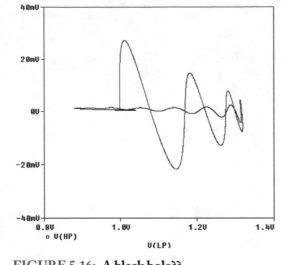

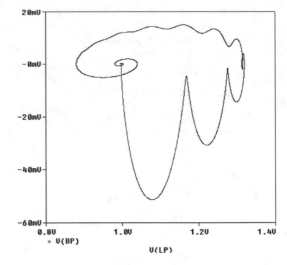

FIGURE 5.16: A black hole??

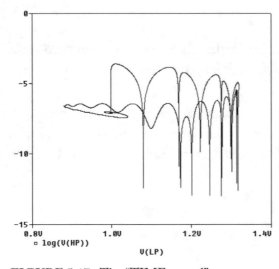

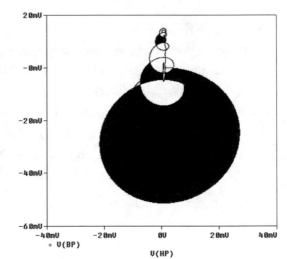

FIGURE 5.17: The "TIME tunnel"

see the effect on the displayed signals [Ref: 5]. The bandpass out on the y-axis versus the low-pass output on the x-axis is shown in Fig. 5.16.

The y-axis (the bandpass output) versus the high-pass output on the x-axis is plotted in Fig. 5.17.

Try sweeping a parameter and changing the x-axis. You are not permitted to use more than one variable in the x-axis in Probe, but extract one plot using the command @ from the **Probe/Trace** ADD list in Fig. 5.18.

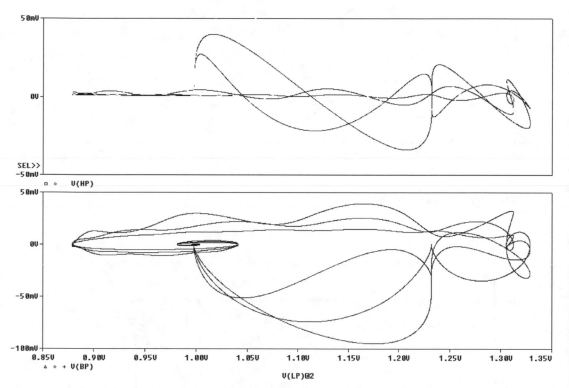

FIGURE 5.18: Select waveform with @ function

FIGURE 5.19: The result of sweeping Rvar1 and selecting the second value

This produces the display shown in Fig. 5.19.

Changing the x-axis to the high-pass output produces the result in Fig. 5.20.

(5) Apply the bandstop frequency transformation $\$ = Bs/(s^2 + \omega_o^2)$ to a Butterworth first-order loss function and invert for a bandstop transfer function. B is the -3 dB bandwidth $= \omega_{p2} - \omega_{p1} = 4574$, and $\omega_0 = 2\pi \times 10^3$. The loss function is $A(s) = [\frac{4574s}{s^2+(2\pi \times 10^3)^2}] + 1 \Rightarrow H(s) = \frac{1}{A(s)} = \frac{s^2+(2\pi \times 10^3)^2}{s^2+4574s+(2\pi \times 10^3)^2}$. Use a Laplace part to produce the ac response. Investigate the relationship $(1 - \text{bandpass transfer function})$, to produce

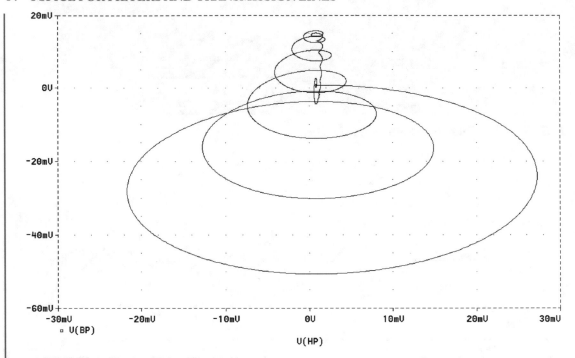

FIGURE 5.20: Shades of the 60's

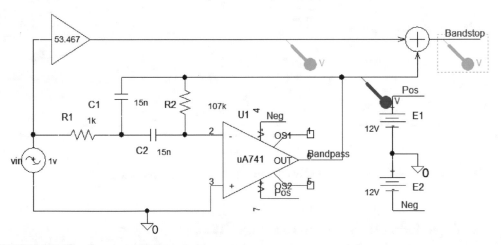

FIGURE 5.21: Bandstop active filter

a bandstop filter as shown in Fig. 5.21. The output from the signal generator is increased to normalize the two outputs. Since the IGMF is an inverting type, then we add, rather than subtract, the two outputs.

The response for the two filter types is shown in Fig. 5.22.

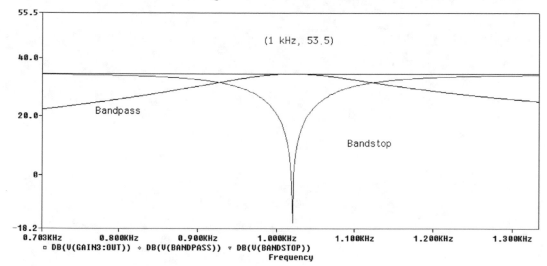

FIGURE 5.22: Bandstop and bandpass response

CHAPTER 6

Switched-Capacitor Filter Circuits

6.1 ELECTRONIC INTEGRATION

An integrator behaves in a similar manner to the flywheel in a car engine, which evens out the mechanical phases of the system. A low-pass filter is an integrator and behaves similarly by integrating, or smoothing out, fast transients in a signal. However, an input square wave applied to a *CR* integrator should have a period much less than circuit time constant $\tau = CR$, in order for integration to occur. This constraint limits the output voltage to a value much less than the input signal amplitude but can be overcome using an active integrator. Select the **ABM** integrator in Fig. 6.1 and set the gain to 10,000. The output wire segment is called **out using the Net Alias** icon. Set **Run to time** to 1 m, and **Maximum step size** (leave blank but not zero). Press **F11** to simulate.

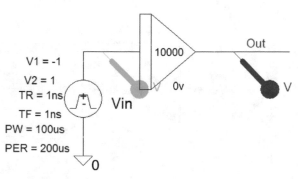

FIGURE 6.1: ABM integrator

The triangular integrated output is shown in Fig. 6.2, when we apply a square wave whose value is ±1. If we have a square wave 0 to 1 V, then the integrated output will continue to rise with time.

6.1.1 Active Integration

The active integrator in Fig. 6.3 has a time constant approximately ten times the period of the input pulse waveform.

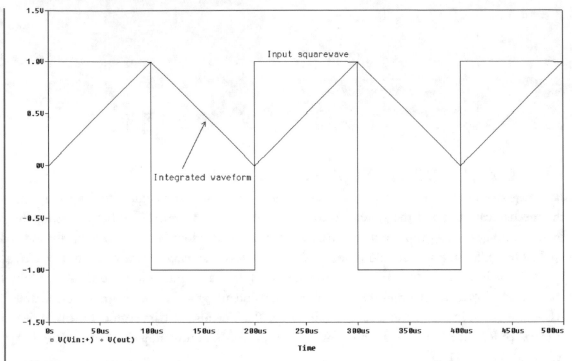

FIGURE 6.2: ABM output for gain set to 10,000

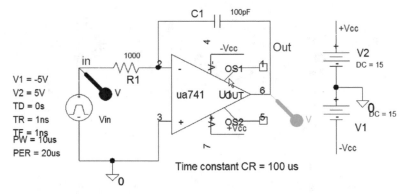

FIGURE 6.3: Active integrator

The input signal is a VPULSE part with parameters as shown. Set the **Output File Options/Print values in the output file** to 0.1 ms, **Run to time** to 1 m, and **Maximum step size** (leave blank), tick **Skip initial transient,** and simulate. Fig. 6.4 shows the output voltage saturating to the −15 V power supply rail voltage. Select **Plot/Unsynchronize X Axis** to observe the signals with a different time scale on the top plot.

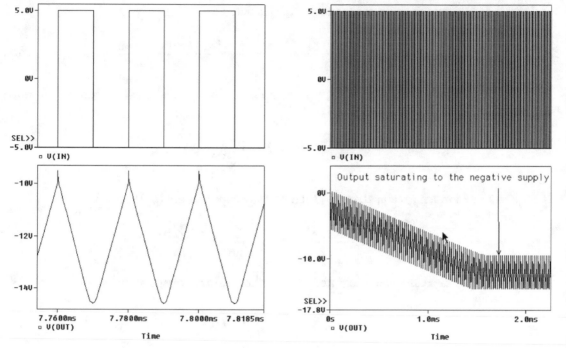

FIGURE 6.4: The output voltage saturating at the rail voltage

6.1.2   Active Leaky Integrator

The output is prevented from saturating due to opamp offsets by placing a large resistance across the capacitance. This is called a leaky integrator as shown in Fig. 6.5. To find the relationship between the input and output voltages, let V_x be the node voltage at the junction of R_1, C_1

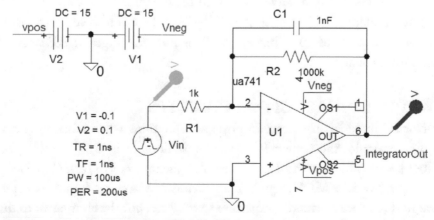

FIGURE 6.5: A leaky integrator

and R_2. The currents in these components are calculated as

$$i_1 + i_2 + i_3 = \frac{V_{in} - V_x}{R_1} + C_1 \frac{d(V_x - V_{out})}{dt} + \frac{V_x - V_{out}}{R_2}. \tag{6.1}$$

If the gain is very large (100 k, for example), then a virtual earth occurs at the node voltage Vx.

$$\frac{V_{in}}{R_1} - C_1 \frac{d(V_{out})}{dt} - \frac{V_{out}}{R_2} = 0. \tag{6.2}$$

If R_2 is assumed much greater than R_1, then we may approximate (6.2) as

$$\frac{V_{in}}{R_1} - C_1 \frac{d(V_{out})}{dt} = 0 \Rightarrow \frac{V_{in}}{C_1 R_1} = \frac{d(V_{out})}{dt}. \tag{6.3}$$

Integrating (6.3) expresses the integrated saw-toothed output voltage signal as

$$V_{out} = \frac{1}{C_1 R_1} \int V_{in} dt. \tag{6.4}$$

Apply the magnifying tool to a section of the saw-tooth signal and verify the slope of the ramp signal is

$$\frac{\Delta V}{\Delta T} = \frac{V_{in}}{C_1 R_1} \Rightarrow \Delta V = \frac{V_{in} \Delta T}{\tau}. \tag{6.5}$$

From the **Analysis** setup menu, select **Transient** and set the following parameters: **Output File Options/Print values in the output file** = 100 ns, **Run to time** = 10 m, and **Maximum step size** = 0.1 μ. Press **F11** to observe the waveforms shown in Fig. 6.6.

Substituting values into (6.5) yields $V = \frac{0.1V 100u}{1u} = 10\ V$. However, because of the ua741 limitations, the output is not 10 V but 8.5 V. Substitute the superior AD8041 opamp and measure the output voltage.

6.2 SWITCHED-CAPACITOR CIRCUITS

Semiconductor IC materials have certain physical limitations when fabricating resistors and capacitors. For example, the largest IC capacitor achievable is $C_{max} = 100$ pF. A MOS integrated low-pass RC filter, with a 1 kHz cut-off frequency ($f_p = 1/(2\pi C_{max} R) = 1$ kHz $\Rightarrow R = 1/(2\pi C_{max} f_p) = 1.59$ MΩ), requires manufacturing a large resistance. A large resistance requires a large IC surface area and is expensive to produce, but the alternative to this is to use a switched capacitor design.

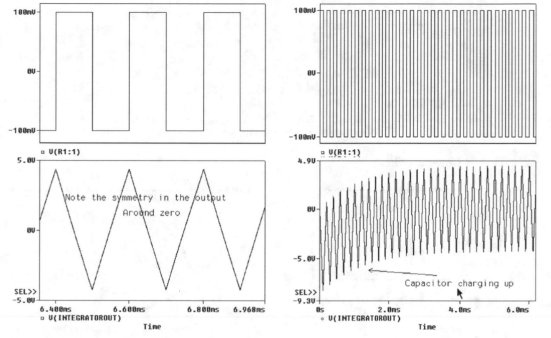

FIGURE 6.6: Integrator output

6.2.1 Switched-Capacitor Low-Pass Filter Step Response

What we are doing here is to replace a resistance with a capacitor that is charged and discharged at a high rate, thus making it appear like a resistance. To understand this circuit, consider a resistance connected between two voltage sources. According to Ohm's law, the current in a resistance is

$$i = \frac{V_1 - V_2}{R_1} = \frac{\Delta V}{R_1}.$$
(6.6)

Fig. 6.7 shows two low-pass filters: The second filter is a conventional *LPF*, and the first filter shows how the resistance is replaced with two other capacitors that are switched on and off in sequence. We may compare the transient behavior of each filter by observing the output from each circuit. By opening and closing the two switches, we cause a small current pulse to flow, thus charging $C1$ momentarily. This charge is then passed to the second capacitor when the first switch is opened and the second switch closed. Current flows through S1 and S2 in sharp pulses each time one of the switches is closed. An average current flows if the switches are opened and closed over a regular time interval $\Delta t = T = 1/f_{clk}$. The voltage on the capacitor $C1$ then changes from $V1$ to $V2$, by a charge flowing from C_1, so the current is

$$i = \frac{\Delta Q}{\Delta t} = \frac{(V_2 - V_1)C_1}{T} = \Delta V C_1 f_{CLK}.$$
(6.7)

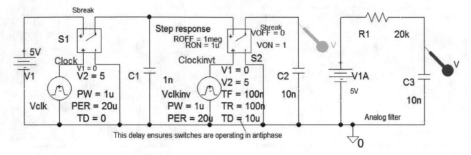

FIGURE 6.7: Low-pass filter

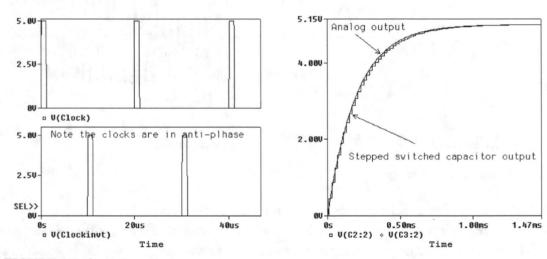

FIGURE 6.8: Anti-phase clocks

Comparing (6.6) to (6.7) implies that

$$R_1 = \frac{\Delta V}{i} = \frac{1}{C_1 f_{clk}}. \tag{6.8}$$

S1 and S2 switches are operated by two out-of-step clocks, whose waveforms are shown in Fig. 6.8 (the second clock has a delay of 10 μs (**TD** = 10 μ).

We compare the step responses of the switched capacitor LPF circuit to the conventional low-pass *CR* circuit response as shown in the right panel of Fig. 6.8. The responses look similar but the switched-capacitor output voltage appears as an exponentially increasing staircase waveform due to the current pulses flowing between the capacitors. To plot the frequency response from the step response as in Fig. 6.9, reset the **Transient Setup** parameter: **Run to time** = 100 ms. Press the **FFT** icon, and select between the *x*-axis numbers to change the settings. Select **User Defined,** and enter 20 Hz to 10 kHz and **Logarithmic**. Select the variables at the bottom and enter 20*log10(V(C3:2)) and 20*log10(V(C3:2)) in the **Trace Expression**

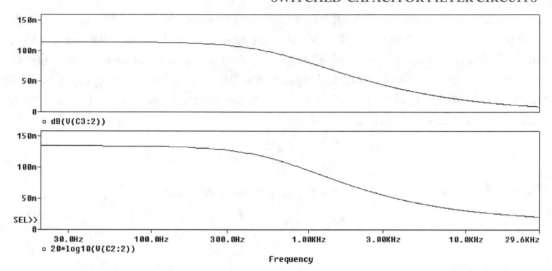

FIGURE 6.9: Frequency response for the two circuits

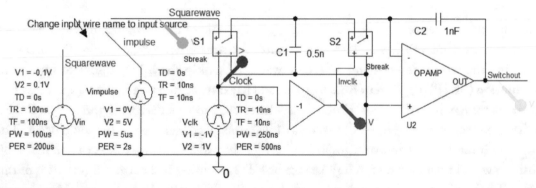

FIGURE 6.10: Switched-capacitor filter

box to plot the response in (or use the dB() function available in the **TRACE** list). Separate the plots using **alt PP**. Set the y-axis to those shown. An error exists at the start of the plot but is reduced by increasing the **Run to time** Transient parameter.

6.2.2   Switched Capacitor Integrator

The resistance in a normal integrator (a LPF) is replaced by a capacitor that is switched on and off electronically as discussed in section (6.2.1). The switched-capacitor filter in Fig. 6.10 uses a 2 MHz clock (frequency f_c) to control the **SBREAK** switch parts, S1 and S2 (from the **Breakout** library). An inverter at the output of the master clock ensures the switches operate in antiphase to each other. $C1$ charges to the input voltage and then transfers its charge to C_2 each time the switch is operated. In a normal integrator, current flows through a feedback capacitor C_2 and is equal to VIN/R (the circuit time constant is RC_2). At the end of half a clock period, $T/2$, C_1 charges up to VIN when the switch S1 is closed and S2 opened. When the clock changes state, S1 opens and S2 closes and the charge on C_1 is transferred to C_2. The amount of charge transferred from the input to the junction at the inverting input of the op amp, during a complete clock period, is $VIN\,C_1$. Since current is the rate of change of charge during a specific time interval, then we write

$$i = \frac{\Delta Q}{\Delta t} = \frac{V_{in}C_1}{T} = V_{in}C_1 f_{clk}. \qquad (6.9)$$

The effective resistance from V_{in} to the negative input is therefore

$$R = \frac{V_{in}}{i} = \frac{1}{C_1 f_{clk}}. \qquad (6.10)$$

The time constant is

$$\tau = RC_2 = \frac{C_2}{C_1 f_{clk}}. \qquad (6.11)$$

The integrator time constant is thus dependent on the ratio of the two capacitors and the clock frequency. To explore the step response, connect the square wave generator to the circuit by changing the wire segment name to square wave. Tick **Skip the initial transient bias point calculation (SKIPBP), Run to time** to 10 m, and **Maximum step size to** 0.1 μ, and press **F11** to simulate. A word of caution about the switches used in this design: If the switch **ROFF** resistance is set to a very large value, such as 1 TΩ, then a convergence error could occur. The lower limit for currents and voltages in PSpice is of the order of $\exp(-10)$. A switch in the off position would then have a switch resistance of 1 TΩ, resulting in a current flow that is of this order. To fix this select the switch part and change **ROFF** to **10MEG** and not 1 M which is a milliohm. Highlight the switch, select **Edit Parameter** and make the change.

This circuit creates a series of voltage steps at the output, with a value $-VinC_1/C_2$. Make C_1 approximately 10 times C_2 to produce a negative ramp with slope:

$$\frac{\Delta V}{\Delta T} = -\frac{V_{in} f_{clk} C_1}{C_2}. \qquad (6.12)$$

Comparing (6.12) to (6.8), we see the time constant is replaced by $C_2/f_{clk}C_1$, so the cut-off frequency is therefore

$$f_c = \frac{f_{clk}C_1}{2\pi C_2} = 15.915 \text{ kHz}. \qquad (6.13)$$

In Fig. 6.11, the "out of step" clocks were unsynchronized by selecting **Probe/Plot** in order to display a few cycles, but with the other signal time axis unchanged.

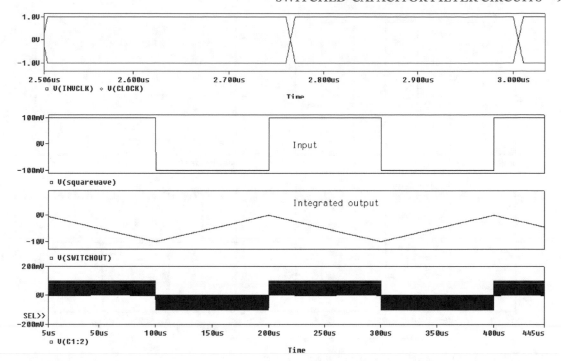

FIGURE 6.11: Switched-capacitor filter waveforms

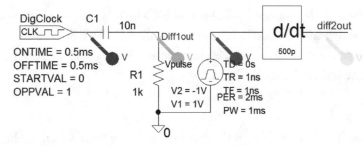

FIGURE 6.12: CR differentiator

A log command file is available from the **Probe/File/Run** menu in probe, and is called **Fig.6-011.cmd.** This separates the waveforms automatically when selected. To apply the impulse to the circuit, select the input wise segment, rename it to impulse and repeat the transient analysis. From **Probe**, use the FFT icon to get the frequency response.

6.3 ELECTRONIC DIFFERENTIATION

A passive differentiator (a high-pass filter) is shown in Fig. 6.12. Differentiating a square wave signal is a useful technique for obtaining narrow pulses that could be used to close an electronic switch at a certain time for example.

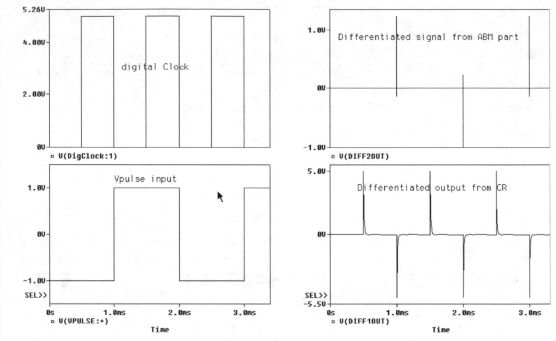

FIGURE 6.13: Differentiating a square wave

An impulse signal is produced when the **pulse width is much less than the circuit time constant**. There is also an **ABM** differentiator part called **DIFFER** as shown with the gain set to 500 p. Set **Run to time** to 10 m and **Maximum step size** (blank). Press **F11** to simulate and produce the narrow pulses shown in Fig. 6.13

6.4   EXERCISES

(1) Percentage Tilt is a measure of the droop in a square wave when applied to a high-pass circuit. It is useful for assessing the low-frequency performance of amplifiers. Copy the differentiator schematic in Fig. 6.12 but increase the capacitor to 5 μF. Set the **Run to time** to 10 ms to allow for transient effects at the beginning of the response and measure the Percentage Tilt from Fig. 6.14. Percentage Tilt is defined as

$$\%\text{Tilt} = \frac{V_1 - V_2}{V_1}100 = \frac{V_1 - V_1 e^{-t/CR}}{V_1}100 \approx \frac{100t}{CR} \approx 200\pi f_c t = 9\ \%$$

where the time substituted into the expression is $t = (9\text{ ms} - 8.5\text{ ms}) = 0.5\text{ ms}$. The exponential voltage decrease is approximated as t/CR (Note: $ex = 1 + x + (x2)/2! + (x3)/3 \cong 1 + x$).

(2) Investigate the active differentiator in Fig. 6.15.

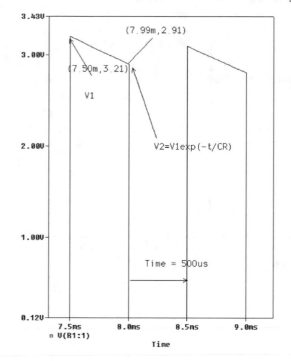
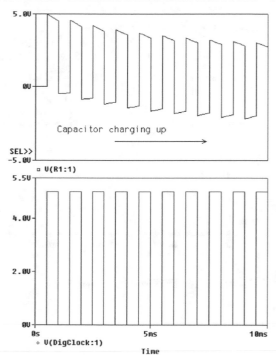

FIGURE 6.14: Measuring percentage tilt

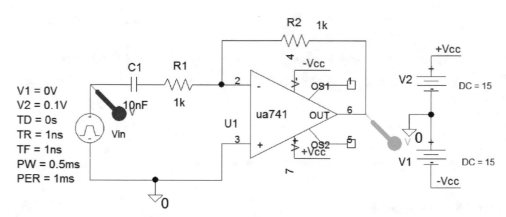

FIGURE 6.15: Active differentiator

The transfer function is $-R_2/(R_1 + j1/\omega C)$. The transfer function for equal value resistors reduces to a passive differentiator transfer function $-R/(R + j1/\omega C)$. The output waveforms are shown in Fig. 6.16. Simulate with, and without, R_1.

(3) Investigate the combined integrator differentiator in Fig. 6.17.

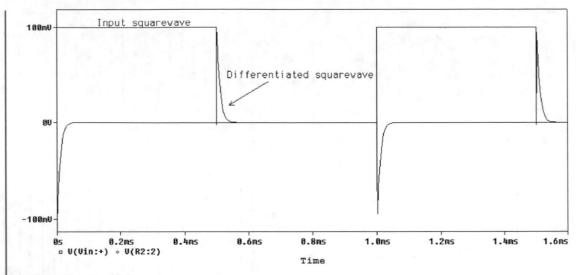

FIGURE 6.16: Differentiator waveforms

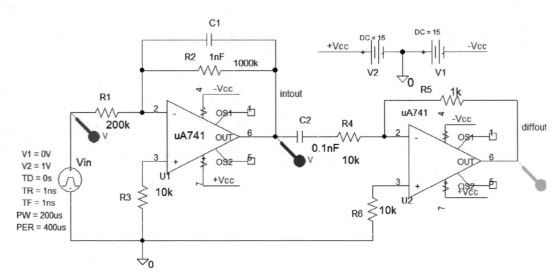

FIGURE 6.17: Integrator differentiator

CHAPTER 7

Two-Port Networks and Transmission Lines

7.1 TWO-PORT NETWORKS

A two-port network has a pair of input and output terminals, as shown in Fig. 7.1. Examples of two-port networks are: Transformers, T-networks, series-tuned LCR circuits, etc. We may classify these as symmetrical or asymmetrical and balanced or unbalanced networks. A network is symmetrical when the input and output terminals are interchangeable without changing the electrical properties of the network. A Tee network with equal series arms is an example of a symmetrical network but is asymmetrical if the series arms are unequal. Further classification is a balanced network where the two input arms (i.e., top and bottom) contain the same elements.

Two-port network analysis uses the input and output currents I_1, I_2, and voltages V_1, V_2 to generate a unique set of equations. Taking the input current, the input voltage, or combinations of currents and voltages as the independent variables, we generate *six* different sets of parameters. For example, if we consider the input and output currents as the independent variables, we generate z-parameters. If we take the input and output voltages as the independent variables, we generate y-parameters. Another set of parameters, the hybrid (mixed Z and Y), or h-parameters, makes the input current and output voltage the independent variables and are used in transistor equivalent circuits.

7.2 THE Z-PARAMETERS

The z-parameters can be used to analyze passive and active networks. The four z-parameters have dimensions of impedance, and, in general, are complex. We define these parameters by opening the input or output terminals (this makes the currents in each circuit zero), hence the name open-circuit. I_1 and I_2 are the independent current variables and generate the following z-parameter equations:

$$E_1 = I_1 Z_{11} + I_2 Z_{12}$$
$$E_2 = I_1 Z_{21} + I_2 Z_{22}.$$

$$(7.1)$$

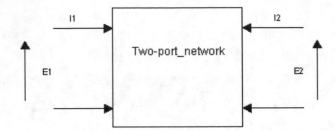

FIGURE 7.1: A two-port network

Open-circuiting the output to define the output impedance (the driving point impedance):

$$Z_{11} = \frac{E_1}{I_1}\bigg|_{I_2=0}. \tag{7.2}$$

An open circuit at the input defines the reverse transfer impedance

$$Z_{12} = \frac{E_1}{I_2}\bigg|_{I_1=0}. \tag{7.3}$$

The output impedance with the input open-circuited is

$$Z_{22} = \frac{E_2}{I_2}\bigg|_{I_1=0}. \tag{7.4}$$

The forward transfer impedance, with the output open-circuited, is

$$Z_{21} = \frac{E_2}{I_1}\bigg|_{I_2=0}. \tag{7.5}$$

The z-parameters are calculated for each parameter and we may then solve for currents and voltages in the circuit.

From the **Analysis Setup**, select **AC Sweep** and **Linear**, **Point/Decade** = 10001, **Start Frequency** = 1k, and **End Frequency** = 10k. Press F11 to produce the response shown in Fig. 7.2.

7.2.1   Z-Equivalent Circuit

We may simulate equivalent circuits such as ac h-parameter equivalents circuit for transistors using ABM parts. Parts such as the voltage-controlled voltage source (VCVS) called an **E** part, a current-controlled voltage source called a **H** part, a voltage-controlled current source called a **G** part, and a current-controlled current source called an **F** part. The z-equivalent is simulated using **ABM** parts as shown in Fig 7.3. The current-controlling voltage sources parameter wires are connected using the **Net alias icon** to name the wire segment.

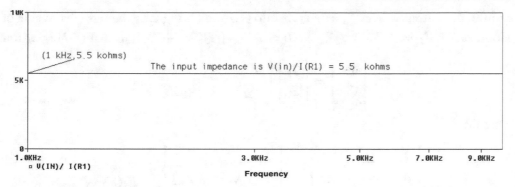

FIGURE 7.2: The input impedence Z in

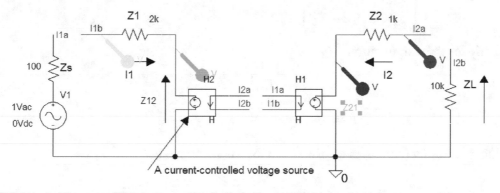

FIGURE 7.3: The z-equivalent circuit using H-parts

Expressions for the current gain, voltage gain, power gain, input impedance, and output impedance can be expressed in terms of the z-parameters. Measure all the z-parameters by substituting short circuits, i.e., 100 meg.

7.2.2 Current Gain

Applying equations (7.2) to (7.5) to the schematic shown in Fig. 7.1, yields: Z11 = R1 + R3, Z12 = Z21 = R3, and Z22 = R2 + R3. From Kirchhoff's first law to the circuit, gives the input and output equations as:

$$E_1 = I_1 Z_{11} + I_2 Z_{12}$$
$$E_2 = I_1 Z_{21} + I_2 Z_{22}. \tag{7.6}$$

The current gain is I_2/I_1, so we must solve for I_1 and I_2 separately. Write equations (7.6) in a matrix form as

$$\begin{bmatrix} E_1 \\ E_2 \end{bmatrix} = \begin{bmatrix} I_1 \\ I_2 \end{bmatrix} \begin{bmatrix} Z_{11} & Z_{12} \\ Z_{21} & Z_{22} \end{bmatrix}. \tag{7.7}$$

Substitute the output voltage $E_2 = -I_2 Z_L$ into (7.7) and bring it to the right-hand side of the equation as $-I_2 Z_L = I_1 Z_{21} + I_2 Z_{22} \Rightarrow 0 = I_1 Z_{21} + I_2(Z_{22} + Z_L)$, so that (7.7) is written as

$$\begin{bmatrix} E_1 \\ 0 \end{bmatrix} = \begin{bmatrix} I_1 \\ I_2 \end{bmatrix} \begin{bmatrix} Z_{11} & Z_{12} \\ Z_{21} & (Z_{22} + Z_L) \end{bmatrix}. \tag{7.8}$$

Solving for I_1

$$I_1 = \frac{\begin{bmatrix} E_1 & Z_{12} \\ 0 & Z_{22} + Z_L \end{bmatrix}}{\begin{bmatrix} Z_{11} & Z_{12} \\ Z_{21} & Z_{22} + Z_L \end{bmatrix}} = \frac{\begin{bmatrix} E_1 & Z_{12} \\ 0 & (Z_{22} + Z_L) \end{bmatrix}}{\Delta Z} = \frac{E_1(Z_{22} + Z_L)}{\Delta Z}. \tag{7.9}$$

Similarly for the output current I_2

$$I_2 = \frac{\begin{bmatrix} Z_{11} & E_1 \\ Z_{21} & 0 \end{bmatrix}}{\Delta Z} \Rightarrow I_2 = \frac{-E_1 Z_{21}}{\Delta Z}. \tag{7.10}$$

Dividing (7.10) by (7.9) gives an expression for the current gain

$$A_i = \frac{I_2}{I_1} = \frac{-E_1 Z_{21}}{\Delta Z} \frac{\Delta Z}{E_1(Z_{22} + Z_L)} = \frac{-Z_{21}}{Z_{22} + Z_L}. \tag{7.11}$$

7.2.3   The Output Impedance

Substituting for the input voltage $E_1 = -I_1 Z_S$ into (7.7) gives $-I_1 Z_S = I_1 Z_{11} + I_2 Z_{12} \Rightarrow 0 = I_1(Z_{11} + Z_S) + I_2 Z_{12}$, so we can write the main equations as

$$\begin{bmatrix} -I_1 Z_S \\ E_2 \end{bmatrix} = \begin{bmatrix} I_2 \\ I_2 \end{bmatrix} \begin{bmatrix} Z_{11} & Z_{12} \\ Z_{21} & Z_{22} \end{bmatrix} \Rightarrow \begin{bmatrix} 0 \\ E_2 \end{bmatrix} = \begin{bmatrix} I_1 \\ I_2 \end{bmatrix} \begin{bmatrix} (Z_{11} + Z_s) & Z_{12} \\ Z_{21} & Z_{22} \end{bmatrix}. \tag{7.12}$$

Solve for the output current I_2

$$I_2 = \frac{\begin{bmatrix} (Z_{11} + Z_S) & 0 \\ Z_{21} & E_2 \end{bmatrix}}{\begin{bmatrix} Z_{11} + Z_S & Z_{12} \\ Z_{21} & Z_{22} \end{bmatrix}} \tag{7.13}$$

$$I_2 = \frac{E_2(Z_{11} + Z_s)}{(Z_{11} + Z_S) Z_{22} - Z_{12} Z_{21}} \Rightarrow \frac{I_2}{E_2} = \frac{Z_{11} + Z_S}{Z_{22}(Z_{11} + Z_S) - Z_{12} Z_{21}}. \tag{7.14}$$

The output impedance is E_2/I_2, so from (7.14), we can write

$$Z_{out} = \frac{E_2}{I_2} = \frac{Z_{22}(Z_{11} + Z_S) - Z_{12}Z_{21}}{Z_{11} + Z_S} = Z_{22} - \frac{Z_{12}Z_{21}}{Z_{11} + Z_S}. \qquad (7.15)$$

7.2.4 Input Impedance

To find an expression for the input impedance, substitute for the output voltage $E_2 = -I_2 Z_L$ into (7.7). The negative sign takes care of the current direction and polarity of E_2 so that the input equation is now $-I_2 Z_L = I_1 Z_{21} + I_2 Z_{22} \Rightarrow 0 = I_1 Z_{21} + I_2(Z_{22} + Z_L)$. The equations are

$$\begin{bmatrix} E_1 \\ -I_2 Z_L \end{bmatrix} = \begin{bmatrix} I_1 \\ I_2 \end{bmatrix} \begin{bmatrix} Z_{11} & Z_{12} \\ Z_{21} & Z_{22} \end{bmatrix} \Rightarrow \begin{bmatrix} E_1 \\ 0 \end{bmatrix} = \begin{bmatrix} I_1 \\ I_2 \end{bmatrix} \begin{bmatrix} Z_{11} & Z_{12} \\ Z_{21} & (Z_{22} + Z_L) \end{bmatrix}. \qquad (7.16)$$

Solve for I_1 using Cramer's Rule.

$$I_1 = \frac{E_1(Z_{22} + Z_L)}{Z_{11}(Z_{22} + Z_L) - Z_{12}Z_{21}} \Rightarrow \frac{I_1}{E_1} = \frac{Z_{22} + Z_L}{Z_{11}(Z_{22} + Z_L) - Z_{12}Z_{21}} \Rightarrow \frac{E_1}{I_1}$$

$$= Z_{11} - \frac{Z_{12}Z_{21}}{Z_{22} + Z_L}. \qquad (7.17)$$

The input impedance is

$$Z_{in} = \frac{E_1}{I_1} = Z_{11} - \frac{Z_{12}Z_{21}}{Z_{22} + Z_L}. \qquad (7.18)$$

Fig. 7.4 shows the steps to measure impedance using an in-built function. Apply this method to measure the input and output impedances for the z-equivalent circuit.

7.2.5 The Voltage Gain

We derive an expression for the voltage gain using expressions for the current gain and input impedance.

$$Av = \frac{V_{out}}{V_{in}} = \frac{E_2}{E_1} = \frac{I_2 Z_L}{I_1 Z_{in}} = A_i \frac{Z_L}{Z_{in}} = \frac{-Z_{21}Z_L}{(Z_{22} + Z_L)Z_{11} - Z_{12}Z_{21}}. \qquad (7.19)$$

7.3 SYMMETRICAL AND UNBALANCED NETWORKS

The characteristic impedance is defined as the impedance looking into one pair of terminals when the other pair of terminals is terminated in the characteristic impedance. Alternatively, it is the input impedance of a network terminated at infinity. Consider a T-network comprised of two series arms $Z_1/2$ each, and a shunt arm $= Z_2$. A two-port network, terminated in

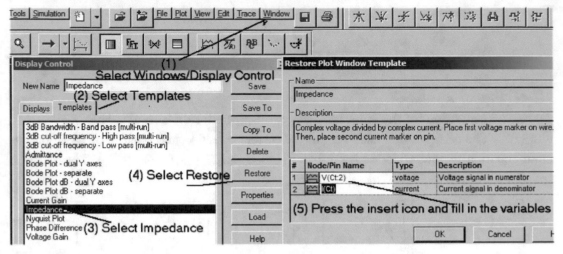

FIGURE 7.4: Using in-built function to measure impedance

the resistance equal to the character impedance, results in an input impedance equal to the characteristic impedance.

7.3.1   Asymmetrical Networks and Image Impedances

A network is **asymmetrical** if we cannot interchange the input and output terminals without affecting the electrical properties of the network. An asymmetrical network thus has two characteristic impedances depending on which side you measure the impedance. For this network the characteristic impedance has a different value when looking from the input or output terminals, so we introduce the concept of **image impedances**. These impedances are defined as the two impedances, which, when one image impedance is connected across the output, the other image impedance is measured across the input.

To simplify the calculations for different networks, we can show the characteristic impedance for a two-port network is the square root of the product of short-circuit and open-circuit impedances.

To prove this, consider the unbalanced symmetrical T-network. The input impedance, when the output is open circuited is $Z_{o/c} = Z_1/2 + Z_2$, and when short circuited is $Z_{s/c} = \frac{Z_1}{2} + \frac{Z_1 Z_2/2}{Z_1/2+Z_2}$. We may now calculate the characteristic impedance using these two expressions as $Z_o = \sqrt{Z_{o/c} Z_{s/c}} = \sqrt{(\frac{(Z_1+Z_2)}{2})(Z_1/2 + \frac{Z_1 Z_2/2}{Z_1/2+Z_2})}$, hence $Z_0^2 = \frac{Z_1}{2}\frac{Z_1}{2} + \frac{Z_1 Z_2}{2} + \frac{Z_1 Z_2}{2} = \frac{Z_1^2}{4} + Z_1 Z_2 \Rightarrow Z_0 = \sqrt{\frac{Z_1^2}{4} + Z_1 Z_2}$. The characteristic impedance for $Z_1 = 50$ Ω and $Z_2 = 100$ Ω is $Z_0 = \sqrt{(50)^2/4 + 50(100)} = 75$ Ω. Substituting into values shows that the input impedance is $Z_{IN} = 25 + 100//100 = 75\Omega \Rightarrow Z_0 = Z_{IN}$. Verify the π network using

$Z_{0\pi} = \sqrt{Z_{o/c}\,Z_{s/c}}$ so $Z_{o/c} = \frac{Z_1(2Z_2)}{Z_1+2Z_2} = \frac{2Z_1Z_2}{Z_1+2Z_2}$ and $Z_{s/c} = \frac{2Z_2(Z_1+2Z_2)}{2Z_2+Z_1+2Z_2} = \frac{2Z_2(Z_1+2Z_2)}{Z_1+4Z_2}$. Thus, the characteristic impedance for the pi network is

$$Z_{0\pi}^2 = \frac{2Z_1Z_2}{Z_1+2Z_2}\left(\frac{2Z_2(Z_1+2Z_2)}{Z_1+4Z_2}\right)$$

$$Z_{0\pi}^2 = \frac{4Z_1Z_2^2}{Z_1+4Z_2}\times\frac{Z_1}{Z_1} = \frac{4Z_1Z_2^2}{Z_1^2+4Z_1Z_2}\frac{\div 4}{\div 4} = Z_{0\pi}^2$$

$$= \frac{Z_1^2Z_2^2}{Z_1^2/4+Z_1Z_2} \Rightarrow Z_{0\pi} = \sqrt{\frac{Z_1^2Z_2^2}{Z_1^2/4+Z_1Z_2}}.$$

The characteristic impedance for the pi networks in terms of the Tee characteristic impedance $Z_{0\pi} = Z_1 Z_2 / Z_{0T}$.

7.4   TWO-PORT RESISTIVE T-ATTENUATORS

Situations arise when a signal is too large and needs to be attenuated but at the same time keep matched conditions such as the "T" and "π" resistive networks attenuating networks. These networks attenuate the signal by a specific amount but keep matched conditions unlike a voltage divider circuit which does not. The T-attenuator in Fig. 7.5 introduces an attenuation of 20 dB (equal to a voltage reduction of ten) but does so keeping matched conditions between the 50 Ω source and load resistances. For example, we might wish to attenuate a signal from a 50 Ω aerial before applying it to a 50 Ω input spectrum analyzer. This resistance value is common in radio frequency communications. The attenuation factor is

$$F = 10^{(-\text{dB}/20)}. \tag{7.20}$$

The resistors of the T-network are expressed in terms of the attenuation factor F. The series element is

$$R_1 = R_L\left((1-F)/(1+F)\right). \tag{7.21}$$

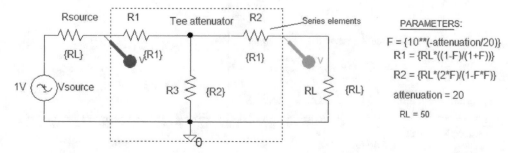

FIGURE 7.5: Resistive attenuator

attenuation	20
F	{10**(-attenuation/20)}
PSpiceOnly	TRUE
R1	{RL*((1-F)/(1+F))}
R2	{RL*(2*F)/(1-F*F)}
RL	50

FIGURE 7.6: Enter expression in the value box

The shunt element is determined as

$$R_2 = R_L(2F)/(1 - F^2). \tag{7.22}$$

For example, if the source and load impedance are 50 Ω, and the network is to introduce 20 dB of attenuation between resistors, then voltage is reduced by ten ($F = 0.1$).

The **PARAM** part has expressions entered instead of actual values as shown in Fig. 7.5. Select the **Param** part and add the extra rows with the parameters as shown in Fig. 7.6.

The attenuated signals are shown in Fig. 7.7.

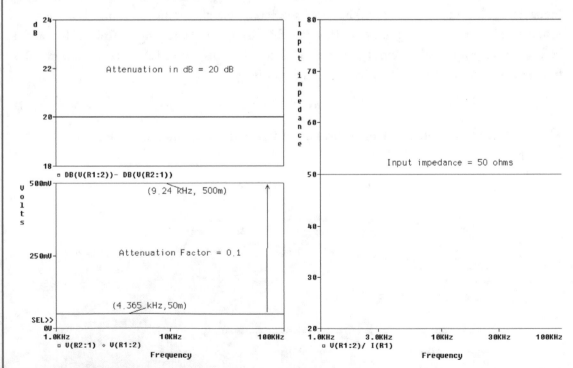

FIGURE 7.7: Frequency response showing the 6 dB attenuation

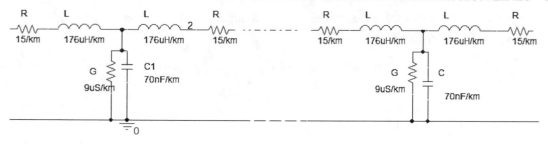

FIGURE 7.8: Lumped transmission line model

7.5   TRANSMISSION LINE MODEL

Transmission lines are formed from two constant cross-sectional area conductors, separated by an insulator (dielectric), and used for telephone lines, guitar leads, TV leads, power leads etc. A transmission line is modeled as a cascaded Tee section of discrete components as shown in Fig. 7.8 and is valid for short lengths of line. In actual transmission lines, these components are distributed evenly along its length.

The primary parameters are per unit length of line defined as

- R: The loop resistance of both conductors,
- G: This represents the leakage current flowing in the dielectric material separating the two conductors,
- C: The capacitance formed from the two conductors and the dielectric, and
- L: The cross-sectional area of the conductor determines the line inductance

PSpice **LEN** parameter defines the unit length and the line component values are per unit length.

7.5.1   Forward and Backward Waves

The voltage at a point l on a transmission line is the sum of the forward and backward waves. $V_1 e^{-(\alpha+j\beta)l} + V_2 e^{(\alpha+j\beta)l}$. The forward wave $V_1 e^{-(\alpha+j\beta)l} = V_1 e^{-\alpha l} e^{-j\beta l}$ is attenuated and phase shifted as it travels along the line. Here, α is the attenuation in **nepers/km**, and the phase change coefficient β, introduces an increasing phase shift along the line. The reflected voltage wave $V_2 e^{(\alpha+j\beta)l} = V_2 e^{\alpha l} e^{j\beta l}$ travels in the opposite direction to the forward wave and is zero if the load impedance equals the characteristic impedance.

7.5.2   Characteristic Impedance

The input impedance of a transmission line is the characteristic impedance Z_0 at any frequency provided the line is terminated in Z_0. An expression for Z_0 is obtained by considering the total voltage divided by the total current. This results in an expression in terms of the primary line parameters R, L, G, and C as

$$Z_0 = \sqrt{\frac{R + j\omega L}{G + j\omega C}}. \tag{7.23}$$

The propagation constant is

$$\gamma = \alpha + j\beta = \sqrt{(R + j\omega L)(G + j\omega C)}. \tag{7.24}$$

The phase shift over a distance $l = \lambda$ meters (one full cycle) is 2π radians expressed as

$$\beta\lambda = 2\pi \Rightarrow \beta = 2\pi/\lambda. \tag{7.25}$$

The velocity of propagation $v = \lambda f$, where f is the frequency of operation (angular frequency ω), hence

$$v = f\frac{2\pi}{\beta} = \frac{\omega}{\beta} m/s.$$

Signal reflection is caused by a mismatched load and the degree of mismatch is measured by the **voltage reflection coefficient**, ρ defined as the ratio of the reflected voltage to the incident voltage. The voltage at any point on the line is the sum of these two voltages, hence, the reflection coefficient is

$$Z_L = Z_0 \left(\frac{1 + \rho}{1 - \rho}\right) \Rightarrow \rho = \frac{Z_L - Z_o}{Z_L + Z_o}.$$

The reflection coefficient for an open and short circuit load resistances is:

Short circuit load, $Z_L = 0$ $\rho = -1$ so $|\rho| = 1$ and $\theta = \pi$,
Open circuit load, $Z_L = \infty$ $\rho = 1$ so $|\rho| = 1$ and $\theta = 0$

7.6   TRANSMISSION LINE TYPES

Transmission line parts in the evaluation version are shown in Fig. 7.9: Loss-less (part name **T**), lossy (part name **Tlossy**), and coupled (part name **T2coupled** and **T3coupled**).

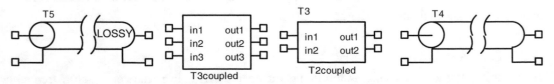

FIGURE 7.9: Transmission line evaluation parts

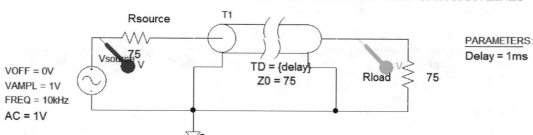

FIGURE 7.10: Lossy transmission line

TD	{delay}
Value	T
Z0	75

FIGURE 7.11: Transmission line parameters

The schematic in Fig. 7.10 shows a 75 Ω lossy **T** part transmission line terminated at the source and load terminals by resistances equal to the characteristic impedance of 75 Ω and a delay TD of 1 ms.

Select the transmission line and set the characteristic impedance to 75 and the delay = 1 ms shown in Fig. 7.11. For matched conditions, the load and source resistances should be equal to the characteristic impedance = 75 Ω.

7.6.1   Lossless Transmission Line Parameters

In many cases we assume the line is loss-less for short transmission lines operating at high frequencies. The propagation constant for a loss-less line is $\gamma = j\beta$, since $\alpha = 0$. Using the relationship $\tanh(j\theta) = j\tan(\theta)$, we express the sending impedance (generator end) as

$$Z_{\text{in}} = Z_0 \left[\frac{Z_L + jZ_0 \tan \beta l}{1 + jZ_L \tan \beta l} \right]. \tag{7.26}$$

The parameters for a loss-less transmission line are shown in Table 7.1.

A lossless transmission line **T** part has parameters defined in terms of a delay in seconds, or a frequency **F** in Hz with a relative wavelength **NL**. (For the default wavelength 0.25, the frequency is the quarter-wave frequency.) When observing time waveforms, the transient step size is limited to one half the delay value specified.

TABLE 7.1: Loss-Less Transmission Line Parameters

SYMBOL	PARAMETER	UNIT NAME	EXAMPLE
ZO	Characteristic impedance	Ohm	75
TD	Transmission delay	Seconds	0.1
F	Frequency for NL	Hertz	1e6
NL	Relative wavelength	Default = (0.25)	0.125

TABLE 7.2: Lossy Transmission Line Parameters

PARAMETER	EXPLANATION	UNIT NAME	SAMPLE
LEN	Electrical length	Meter	1 km
R	Resistance/LEN	Ω/LEN	15 Ω/km
L	Inductance/LEN	Henry/LEN	0.175 mH/km
G	Conductance/LEN	Siemens/LEN	9 uS/km
C	Capacitance/LEN	Farads/LEN	70 nF/km

7.7   LOSSY TRANSMISSION LINE PARAMETERS

Lossy coax transmission line primary line constants are shown in Table 7.2.

The characteristic impedance at $f = 2$ MHz is calculated by substituting the line constants.

$$Z_0 = \sqrt{\frac{R + j\omega L}{G + j\omega C}} = \sqrt{\frac{(15 + j2\pi 2 \times 10^6 \times 175 \times 10^{-6})}{(9 \times 10^{-8} + j2\pi 2 \times 10^6 \times 70 \times 10^{-9})}} = 50 - j0.34 \ \Omega.$$

Draw the schematic in Fig. 7.12 using the **TLOSSY** part and carry out an AC analysis from 0.10 Hz to 100 kHz. The length of the transmission line is 1 km so **LEN** is set to 1 (primary line parameters are normally quoted per km). The propagation constant is

$$\gamma = \alpha + j\beta = \sqrt{(R + j\omega L)(G + j\omega C)}$$
$$= \sqrt{(15 + j2\pi 2 \times 10^6 \times 190 \times 10^{-6})(9 \times 10^{-6} + j2\pi \times 2 \times 10^6 \times 70 \times 10^{-9})}$$
$$= 0.1442 + j46.$$

The real part is measured in Nepers/km, or $\ln(\alpha)$.

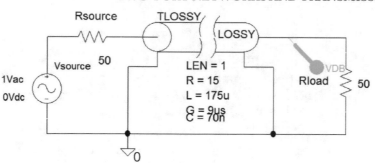

FIGURE 7.12: A 1 km length of lossy transmission line

RL	15
C	70n
G	9u
L	175u
LEN	1

FIGURE 7.13: Tlossy parameters

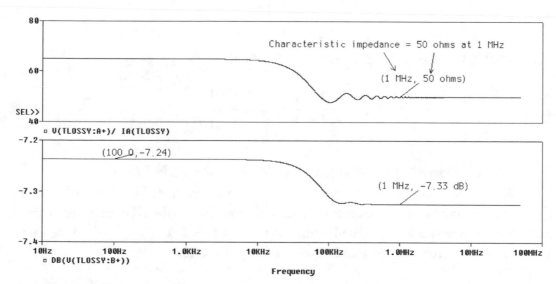

FIGURE 7.14: Plot of the input impedance and frequency response

The **TLOSSY** parameters are shown in Fig. 7.13.

The Probe plot in Fig 7.14 shows the transmission line input impedance plotted by selecting **Trace/Add Trace** (or the Insert button) and dividing the input line voltage by the input line current, i.e., **V(TLOSSY:A+)/ IA(TLOSSY)**.

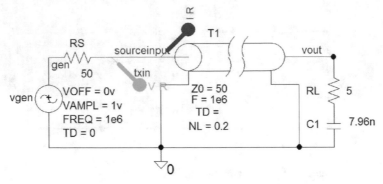

FIGURE 7.15: Problem 1

7.8   VOLTAGE STANDING WAVE RATIO (VSWR)

The maximum and minimum voltages, V_{max} and V_{min}, are measured using a detector along a slotted line (a special line for measuring standing waves). The ratio of the two voltages defines the voltage standing wave ration **VSWR**:

$$S = \frac{V_{max}}{V_{min}}. \tag{7.27}$$

The *VSWR* has a range $1 \leq S \leq \infty$ and it can be shown that

$$|\rho| = \frac{S-1}{S+1} \Rightarrow S = \frac{1+|\rho|}{1-|\rho|}. \tag{7.28}$$

7.9   INPUT IMPEDENCE OF A TRANSMISSION LINE

The 50 Ω transmission line (part name **T**) in Fig. 7.15 is terminated in an impedance $Z_L = 5 - j20\,\Omega$. Determine the impedance 0.2 λ from the load (this is the **NL** parameter in the line part). Since the capacitance is related to the reactance as $C = 1/2\pi X_c f$, the load capacitance is $C = 1/2\pi 20 \times 10^6 = 7.96\,\text{nF}$ (the input frequency is 1 MHz).

7.9.1   Solution

To verify this value in PSpice, we need to set the length of the line to 0.2 λ and plot the real and imaginary parts of the input impedance of the line. Set the **Analysis Setup** from 900 kHz to 10 MHz and simulate. From the Probe output, we plot the imaginary part of the input impedance as $-\textbf{IMG(V(txin)/I(vgen))}$, and the real part of the input impedance as $-\textbf{R(V(txin)/I(vgen))}$, as seen in Fig. 7.16.

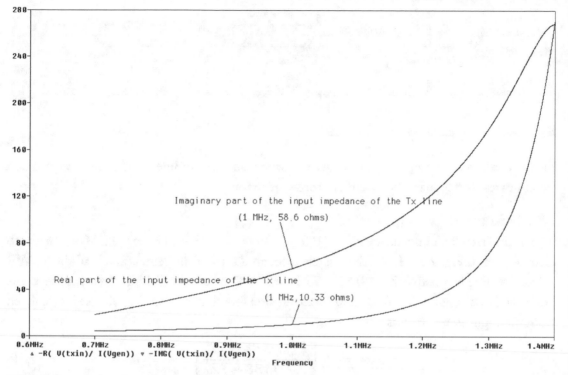

FIGURE 7.16: Plotting real and imaginary parts of input impedance

The values read from Probe are: Real part = 10 Ω and imaginary part is 59 Ω. To solve using a Smith chart we must normalize the load as

$$\frac{Z_L}{Z_0} = \frac{(5 - j20)}{50}\ \Omega = 0.1 - j0.4\ \Omega.$$

Mark this impedance on the Smith chart as point A and draw a line from the center through this point past the outer circle. Read the wavelength value on the outer circle as 0.439 λ. Draw a VSWR circle with radius A representing the impedance (see point D $VSWR = 12$). Point B is (0.439 λ + 0.2 λ − 0.5 λ) = 0.139 λ. (We subtract 0.5λ because the chart is from 0 to 0.5λ.) Draw an arc from 0.439 λ to the point 0.139 λ, and draw a line from this point to the center. The line intersects the drawn circle at 0.2 + $j1.18$. The actual impedance is 50(0.2 + $j1.18$) = 10 + j 59 Ω, which agrees with our PSpice simulation results.

7.10   QUARTER-WAVE TRANSMISSION LINE MATCHING

A load $ZL = 100 − j50\ \Omega$ is connected across a 75 Ω characteristic impedance transmission line shown in Fig 7.17. With the aid of a Smith chart and verifying with PSpice, obtain a value

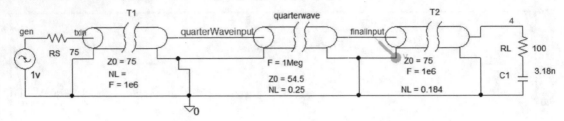

FIGURE 7.17: Quarter–wave transformer

for the characteristic impedance of a quarter-wave transmission line transformer inserted at a point nearest to the load line to achieve correct matching.

7.10.1 Solution

The normalized load impedance $Z_L = (100 - j50)/75 = 1.33 - j0.67$ is point A on the Smith chart shown in Fig. 7.16. Draw the S circle through ZL from the center, and read the VSWR value. The line is resistive $Z_R = 0.53 \times 75\ \Omega = 39.8\ \Omega$ where the VSWR circle intersects the horizontal axis at point $B = 0.53$. A $\lambda/4$ transmission line characteristic impedance is calculated from the input impedance as

$$Z_{in} = Z_0' \left[\frac{Z_L/\tan \beta\ell + j Z_0'}{Z_0'/\tan \beta\ell + j Z_L} \right]. \tag{7.29}$$

The $\lambda/4$ line input impedance is the characteristic impedance of the original line and the total phase is

$$\beta l = (2\pi/\lambda)(\lambda/4) = \pi/2. \tag{7.30}$$

Hence

$$\tan \beta l = \infty. \tag{7.31}$$

The relationship between the input impedance, the characteristic impedance, and the impedance Z_L is

$$Z_{in} = Z_0 = Z_0' \left[\frac{0 + j Z_0'}{0 + j Z_L} \right] = \frac{Z_0'^2}{Z_L} \Rightarrow Z_0' = \sqrt{Z_0 Z_L}. \tag{7.32}$$

The input impedance of a $\lambda/4$ section of line should equal the characteristic impedance of the line to be matched. Substituting $Z_L = Z_R$ into the quarter-wave transformer equation yields the characteristic impedance for the $\lambda/4$ line:

$$Z_0' = \sqrt{Z_0 Z_R} = \sqrt{75 \times 39.8} = 54.5\ \Omega. \tag{7.33}$$

The quarter-wave transmission line is located $0.184\ \lambda$ from the load.

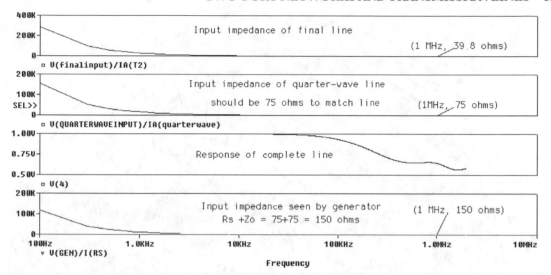

FIGURE 7.18: Quarter-wave transformer matching device

The inserted $\lambda/4$ line parameters are: **NL** = 0.25 and $\mathbf{Z_0}$ = 54.5 Ω. From the **Analysis Setup**, select **AC Sweep** and **Linear**, **Point/Decade** = 10001, **Start Frequency** = 100, and **End Frequency** = 10meg. The load values are calculated as per problem 1. Press **F11** to produce the Probe results in Fig. 7.18.

Determine the characteristic impedance of the inserted line using the Smith chart in Fig. 7.19. What is the distance from the termination, d, where it is to be inserted?

7.11   SINGLE-STUB MATCHING

Terminating a transmission line in a value other than the line characteristic impedance can lead to problems. Fig. 7.20 shows one technique of matching the line using a short-circuit stub as a certain location and whose characteristics are the same as the line to be matched.

To obtain the values for this schematic we first solve using the Smith chart and then back track. The frequency of operation is 100 MHz.

7.11.1 Problem 1

Standing waves on a 50 Ω transmission line had a maximum voltage V_{max} = 10 mV and a minimum voltage V_{min} = 2 mV measured 50.8 mm from the load. If the wavelength is 212 mm at the operating frequency, determine:

- The VSWR and the magnitude of the reflection coefficient,

- The distance between the load and the first minimum (in wavelengths), and,

- The return and mismatch loss in dB.

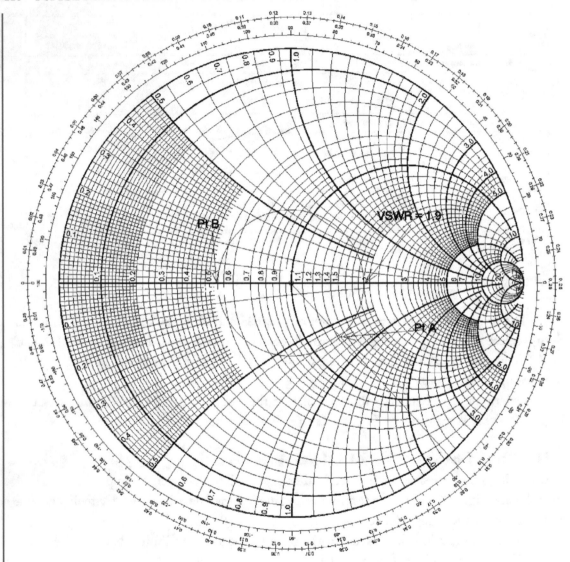

FIGURE 7.19: Quarter-wave transformer

Determine the location and length of a single stub with the same characteristics as the mismatched line to achieve correct matching.

7.11.2 Solution

In Fig. 7.21, rotating around from the load toward the generator will, at some point, result in the impedance of the line having a normalized real part of unity but will have an imaginary part that requires to be cancelled by the stub connected across the line at this point. If a stub inserted

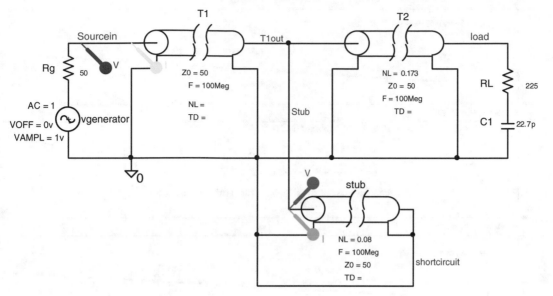

FIGURE 7.20: Stub matching

at this point has an opposite reactance to the reactance of the line, then the net impedance is unity (the reactance parts cancel). The line is then matched on the generator side of the single stub. There are two factors to be determined: the distance from the load to the stub, and the length of the stub itself, which is a short-circuit line in parallel with the line. The voltage standing wave ratio VSWR, and the reflection coefficient is

$$VSWR = \frac{V_{max}}{V_{min}} = \frac{10}{2} = 5$$

$$\rho = \frac{VSWR - 1}{VSWR + 1} = \frac{5 - 1}{5 + 1} = 0.67.$$

The location is

$$D_{min} = \frac{50.8 \text{ mm}}{212 \text{ mm}} = 0.24\lambda.$$

Locate the point $D_{min} = 0.24\,\lambda$, and draw a line from the center of the chart to this point. The *intersection* of the line, and the circle of constant *VSWR*, is the point A, and is the normalized impedance **4.5 − j1.4** Ω. The actual load is $50(4.5 - j1.4\,\Omega) = 225 - j70\,\Omega$. At a frequency of 100 MHz, the reactance of a capacitance of 22.7 pF is 70 Ω.

We work with **admittance** since we use a short circuit stub in parallel with the line. The admittance of the load is located through the center from A, to the **opposite side** where it intersects the *VSWR* circle at **Pt B,** an equal distance from center and is read as $0.2 + j0.07\,S$.

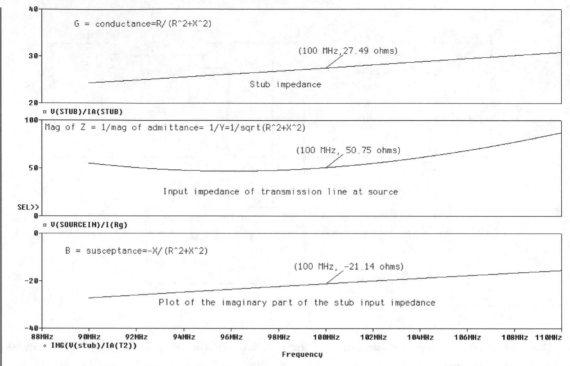

FIGURE 7.21: Single stub matching

On the outer circle read the wavelength measurement as 0.01λ. Travel around the constant *VSWR* circle to **point C** (**y = 1 + j1.8 S),** where it intersects the unity conductance circle. A stub whose admittance is a pure susceptance **ys = −j1.8 S,** is connected across the line, where it cancels out the line susceptance of **j1.8 S.** The stub location is **(0.183 − 0.01)λ = 0.173λ** from the load toward the generator. The susceptance **−j1.8 S** is read at 0.33 λ, i.e., the stub length is **(0.33λ − 0.25λ) = 0.08λ = 0.08*212 mm = 17 mm,** and is placed a distance 0.173λ*212 mm = 36.7 mm from the load. See exercise (3) at the end of the chapter. Transmission line losses due to reflection are determined using a **MACRO**. The magnitude of the standing wave ratio is defined:

$$\rho = \frac{(\mathrm{abs}(Z_L) - Z_0)}{(\mathrm{abs}(Z_L) + Z_0)}. \tag{7.34}$$

The input transmission line nodes are defined as +**A,** and −**A** (normally grounded), and the output nodes +**B** and −**B** (normally grounded). For example, the output node for the second transmission is **VB1(T2),** The load may be in real and imaginary form, so the macro for computing the absolute value of the load is: magZL(RL,XL) = sqrt(RL*RL+XL*XL). The

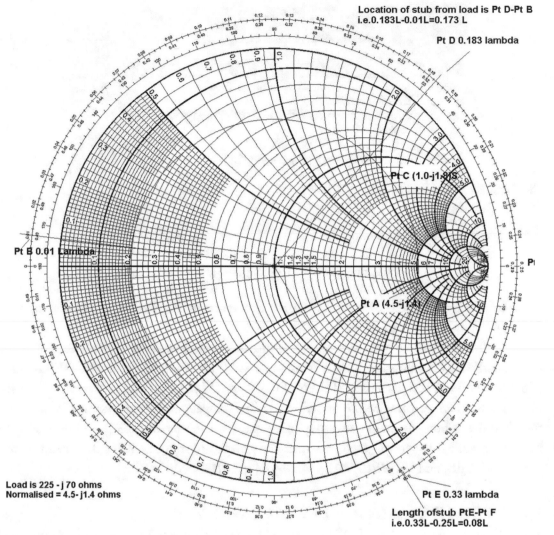

FIGURE 7.22: Single stub matching

macro for calculating the magnitude of the reflection coefficient is given as:

$$\text{mag\_rho}(\text{magZL}, \text{ZO}) = (\text{magZL} - \text{ZO})/(\text{magZL} + \text{ZO}) \qquad (7.35)$$

The mismatch loss ML $= -10\log[1 - \rho^2] = -2.5$ dB,
     The macro for this function is:

$$\text{mismatchloss}(\text{mag\_rho}) = 10 * \log 10(1 - \text{mag\_rho} * \text{mag\_rho}) \qquad (7.36)$$

$$\text{Returnloss} = 20 * \log 10(\text{mrho}) \qquad (7.37)$$

$$\text{VSWR} = (1 + |\rho|)/(1 - |\rho|). \qquad (7.38)$$

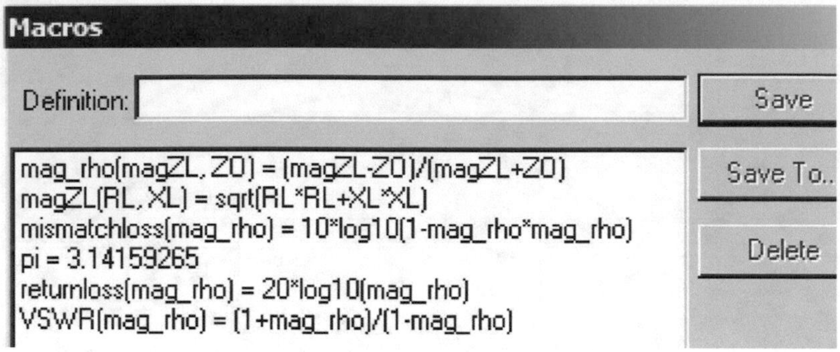

Macros

Definition: [   ]   Save

mag_rho(magZL, Z0) = (magZL-Z0)/(magZL+Z0)
magZL(RL, XL) = sqrt(RL*RL+XL*XL)
mismatchloss(mag_rho) = 10*log10(1-mag_rho*mag_rho)
pi = 3.14159265
returnloss(mag_rho) = 20*log10(mag_rho)
VSWR(mag_rho) = (1+mag_rho)/(1-mag_rho)

Save To..

Delete

FIGURE 7.23: Macro definition

The macro for this function is:

$$\text{VSWR(mag\_rho)} = (1 + \text{mag\_rho})/(1 - \text{mag\_rho}).$$

Selecting **Trace** and **Macro...** from the Probe screen displays the list of macros saved from a previous session in Fig. 7.23.

In the **Definition** box, type in the macro function, and press **Save**. The macros are stored with the file extension **.prb**. For example C:\Pspice\Circuits\chapter02\FIGURE7-026-PSpiceFiles\FIGURE7-026\TRAN\TRAN.prb. If the function is complex with lots of brackets, then you are better off typing it in Notepad, and then pasting it back into the **Definition** box. Examples of using these macros are: magZL(225,70) returns 235 W, mag_rho(ma235,50) returns 0.65, mismatchloss(mag_rho) returns 2.5 dB, returnloss(rho) returns −3.6 dB, and VSWR(0,541) returns 4.8 [ref: 6].

7.12   TIME DOMAIN REFLECTROMETRY

Time domain reflectrometry (TDR) is a technique for locating transmission line faults. Such faults can occur if water gets into the line when it is accidentally cut. The water changes the termination conditions and produces reflections back toward the source. When a transmission line is terminated in a load not equal to the characteristic impedance, a portion of the wave at the load is reflected back to the source. The magnitude of the reflected signal depends on the degree of mismatch and is quantified by the reflection coefficient. The source will not see the reflected signal until *time* = $2T$, where T is the time it takes to travel the transmission line length. The voltage at the load is then the sum of the incident and the reflected signals. A second reflection takes place when the signal reaches the source. This process continues with the signal being reflected from the load back to the source etc. A lattice, or bounce diagram, may be drawn from the results measured.

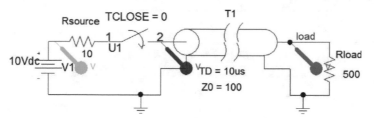

FIGURE 7.24: Creating a bounce diagram

7.12.1 The Lattice/Bounce Diagram

Fig. 7.24 shows the following transmission line parameters: $Z_0 = 100$ Ω and delay $TD = 10$ μs.

A 10 V DC supply with a 10 Ω source resistance is connected to a transmission line of length l meters (delay $= 10$ μ), and characteristic impedance of 50 Ω and is terminated in 500 ohms. T is the amount of time required for a signal to travel the length of the transmission line. The voltage at the transmission line input is

$$\frac{V_{TXin}}{V_{step}} = \frac{Z_0}{R_{source} + Z_0} = \frac{100}{110} = 0.909 \text{ V}. \qquad (7.39)$$

Set the Analysis tab to Analysis type: **Time Domain** (Transient), **Run to time** $= 200$ us, and **Maximum step size** $= 100$ ns, and press F11 to simulate.

Fig. 7.25 shows the reflections from the source and load because of the mismatched impedances at either end.

7.13   TDR FOR FAULT LOCATION ON TRANSMISSION LINES

Faults develop on a transmission line due to water getting into the cable (pressuring the line helps prevent this) and the location of these faults is generally unknown. Some means of identifying the fault location is required and Fig. 7.26 demonstrates TDR is used to locate a fault on a line. Here, we place a "fault" of 10 Ω at a "known" location.

From the diagram in Fig. 7.27, see if you can explain the wave shape signals to "locate" the fault.

7.14   STANDING WAVES ON A TRANSMISSION LINE

Figure 7.28 shows transmission lines terminated with four loads (i) open-circuit, (ii) short-circuit, (iii) 37.5 ohms and 75 ohms.

Investigate the resulting standing waveforms plotted in Fig. 7.29.

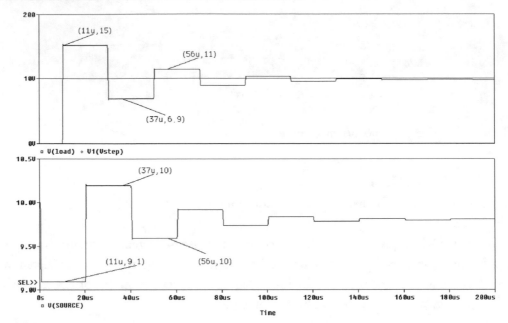

FIGURE 7.25: Bounce diagram

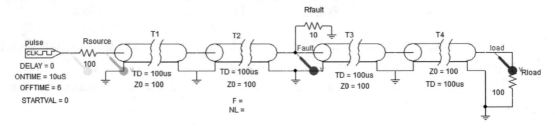

FIGURE 7.26: Fault location

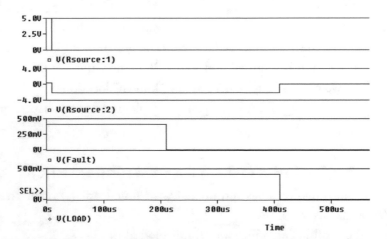

FIGURE 7.27: TDR signals

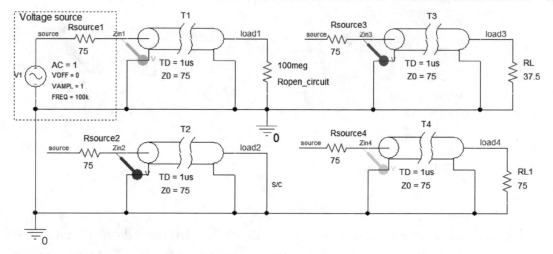

FIGURE 7.28: Transmission line with four different loads

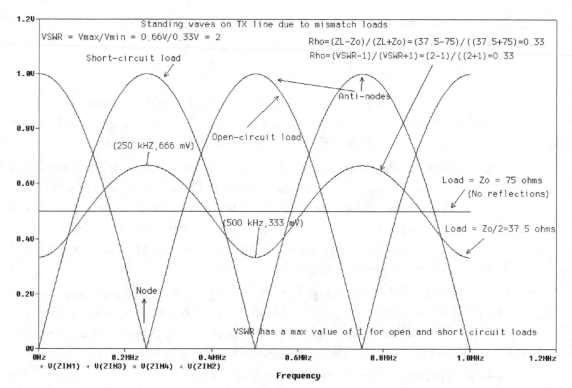

FIGURE 7.29: Transmission line standing waves

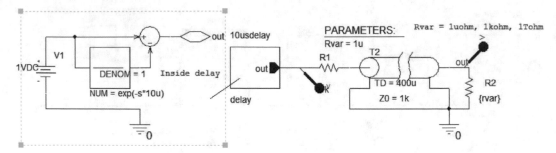

FIGURE 7.30: TDR measurements

7.15   EXERCISES

(1) A 10 km telephone cable, at a frequency of 10 kHz, has the following primary transmission line parameters: $L = 700$ mH/km, $C = 0.05$ μF/km, $R = 28$ Ω/km, and $G = 1$ μS/km. Determine the phase and attenuation constants and calculate the characteristic impedance.

(2) The input impedance of a transmission line, terminated in a resistance $Z_R = 35$ Ω is

$$Z_{in} = Z_0 \left[\frac{Z_R / \tan \beta \ell + j Z_0}{Z_0 / \tan \beta \ell + j Z_R} \right]$$

where β is the phase change coefficient and Z_0 is the characteristic impedance equal to 75 Ω. Determine the input impedance of a section of line whose length is $\lambda/4$ meters. Describe a graphical technique for matching a transmission line to a load.

(3) Define voltage standing wave ratio and voltage reflection coefficient. A transmission line, whose characteristic impedance $Z_0 = 300 + j0$ Ω, has an antenna of impedance $225 - j175$ Ω connected as a load. Matching by means of a single stub connected a distance d meters from the load, is used. Estimate the stub length l in meters, and the distance d, if the operating frequency is $f = 500$ MHz. Assume that the stub is formed from a section of the same air-spaced transmission line.

(4) Investigate the TDR circuit in Fig. 7.30. A unit impulse is created and connected to a transmission line that is open-circuited at the output. A reflection will be detected at the input. Measure the delay. Omit the source resistance and repeat the measurements. Place a short circuit on the output and observe the waveforms.

(5) Investigate the coupled transmission lines shown in Fig. 7.31.

(6) Apply a 1 kHz square wave signal to the schematic in Fig. 7.32 and observe the filtering action.

The smeared output is observed in Fig. 7.33.

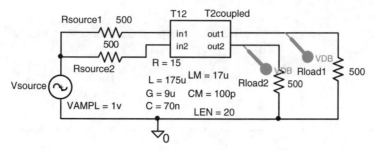

FIGURE 7.31: Coupled lines

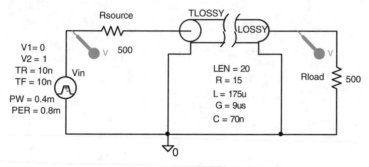

FIGURE 7.32: Transmission line as a low-pass filter

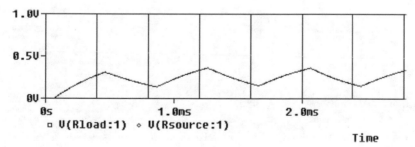

FIGURE 7.33: Smeared output

(7) Design a π attenuator to operate between 300 Ω impedance and introduce 6 dB between source and load. The network design equations are $R_1 = R_L((1 + F)/(1 - F))$, and $R_2 = R_L(1 - F^2)/(2F)$.

(8) The "matched" loss-less transmission line in Fig. 7.34 is a loss-less transmission line whose characteristic impedance is given approximately as:

$$Z_0 \approx \sqrt{L/C} = \sqrt{175 \times 10^{-6}/70 \times 10^{-9}} = 50\,\Omega.$$

Plot the voltage/current waveforms to show the attenuation in dB at the input (or output) is $20\log[50/(50 + 50)] = -6$ dB.

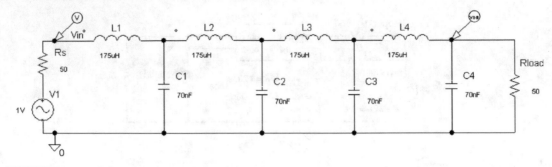

FIGURE 7.34: A loss-less transmission line

CHAPTER 8

Importing and Exporting Speech Signals

8.1   LOGIC GATES

DC supplies are not required when simulating logic circuits because PSpice hides power supply connections to simplify the schematic. However, you may wish to place supplies to investigate power supply decoupling (see PSpice for circuit theory and electronic devices). There are a limited number of signal transitions in the evaluation version, so be careful about component selection. For example, use a **DigClock** part instead of a **VPULSE** generator for square waves, because PSpice has to insert an analog to digital converter on the **VPULSE** if used. Fig. 8.1 shows all input binary combinations applied to two-input NAND (7400) and (7408) AND logic gates terminated in a wire segment. Use the Net alias icon to place names on the wire segments. The results are displayed on the schematic after ticking **Transient** in the **Analysis Setup** menu. To display the logic levels make sure you have a wire segment between the **Hi** (or **Lo)** part, the gate input and outputs, and the output wire segment. Select the DC bias icon **V** to display logic levels. Note: The **HI** and **LO** symbols have been renamed and are placed by selecting the power icon from the right icon toolbar. Carry out the same procedure for the NOR (7402) and XOR (7486) gates with results as shown in Fig. 8.2.

The OR gate (7432) is simulated for all permutations of input signals in Fig. 8.3. Also shown is an OR gate and inverter part to form an XNOR gate as the evaluation version of PSpice does not include an XNOR gate in the library.

To display analog voltage levels rather than logic levels, terminate the output with a resistance. Analog to digital interfaces are automatically inserted between the open collector drivers and the 10 kΩ pull-up resistor.

8.2   BUSSES AND MULTIPLEX CIRCUITS

Table 8.1 shows a 16-to-8 pin multiplexer reduction algorithm to demonstrate the use of a 4-bit digital stimulus **STIM4** which produces four parallel outputs connected to the circuit via

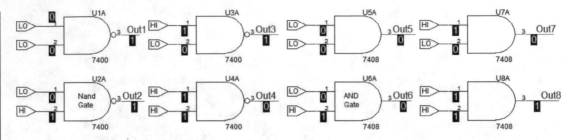

FIGURE 8.1: NAND and AND gates

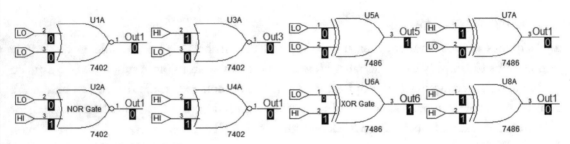

FIGURE 8.2: NOR and XOR gates

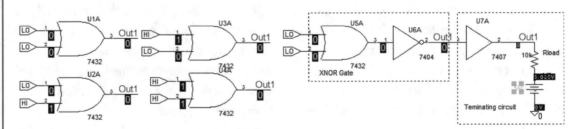

FIGURE 8.3: OR and XNOR gate connection

a wire bus. The wire bus system is a neater wire-connection method and is placed by selecting the icon (or **B**).

Set up the 16 to 8 multiplexing schematic in Fig. 8.4.

To draw a bus, press **B** or use the fifth item from the toolbar icon in Fig. 8.5.

Lclick and move the mouse in the desired bus direction. Select **N1** or press key **N**, to create a net alias (the bus name) and enter **D[0-3]** in the **Alias** name box. This produces a little box which must be dragged to sit on the drawn bus. Press the Esc key to stop placing bus names. This procedure is also the same for the wires connecting to the bus. The generator

TABLE 8.1: 16-to-8 pin multiplexer logic table

DIGIT	D3	D2	D1	D0	OUT	PAIR	PIN	PAIR	TG
0	0	0	0	0	0	D0	Pin0	Same	1
1	0	0	0	1	1	D0			0
2	0	0	1	0	1	D0bar	Pin1	Invert	0
3	0	0	1	1	0	D0bar			0
4	0	1	0	0	0	D0	Pin2	Same	0
5	0	1	0	1	1				0
6	0	1	1	0	1	Hi	Pin3	2×1=hi	0
7	0	1	1	1	1				0
8	1	0	0	0	1	D0bar	Pin4	Invert	0
9	1	0	0	1	0				0
10	1	0	1	0	1	D0bar	Pin5	Invert	0
11	1	0	1	1	0				0
12	1	1	0	0	1	Hi	Pin6	2×1=hi	0
13	1	1	0	1	1				0
14	1	1	1	0	0	D0	Pin7	Same	0
15	1	1	1	1	1				0

has four outputs which must be connected to the rest of the circuit via wires. Each wire is labeled **D0, D1, D2, and D3**. Select and **STIM4** and fill in the parameters as shown in Fig. 8.6.

The multiplexed signals are displayed in Fig. 8.7.

8.3   RANDOM ACCESS MEMORY

Random access memory (RAM) is a temporary store for data signals. One example where we could use RAM is to implement digital reverberation used in the music industry by sampling the voice and storing the binary output in RAM. The data is then read out from the RAM at a short time later and applied to a DAC where it is mixed with the original signal. The schematic in Fig. 8.8 uses the bus system as a means of connecting multilines from source output to the IC inputs. Generator parts **STIM4, STIM8**, and **STIM16** have four, eight, and sixteen parallel

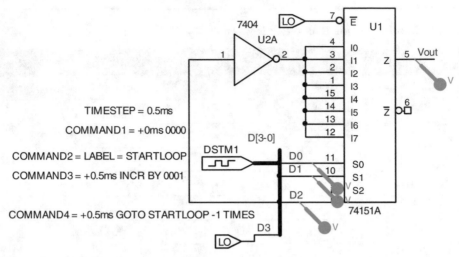

FIGURE 8.4: 16-8 multiplexing

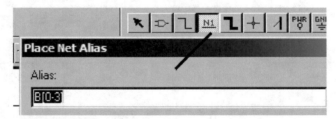

FIGURE 8.5: Placing net aliases

COMMAND1	+0ms 0000
COMMAND2	LABEL = STARTLOOP
COMMAND3	+0.5ms INCR BY 0001
COMMAND4	+0.5ms GOTO STARTLOOP -1 TIMES
TIMESTEP	0.5ms
Value	STIM4
WIDTH	4

FIGURE 8.6: The parallel data signal generator STIM4

outputs for stimulating bus lines in digital circuits and are connected here using the bus system explained previously.

Any unused lines from **STIM4 and STIM16** should be terminated with a LO part. Placing a marker on a bus will display the Probe output signal in hexadecimal format as shown in Fig. 8.9. Digital signals displayed in Probe can sometimes occupy only a small section of the

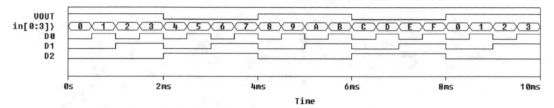

FIGURE 8.7: Signal waveforms

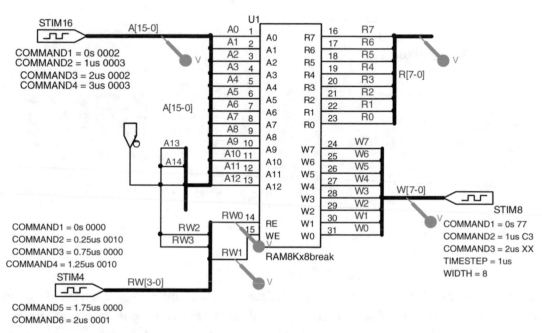

FIGURE 8.8: Investigating RAM basics

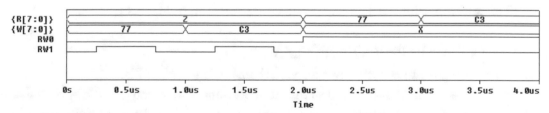

FIGURE 8.9: RAM signals

total display area. To compress any digital graphs in Probe, press the restore icon at the top left hand section of the Probe output. This will cause a small reduction in picture size. When the Probe screen is reduced, place the cursor at the bottom right corner and resize the graph. Repeat for the left hand side of the graph.

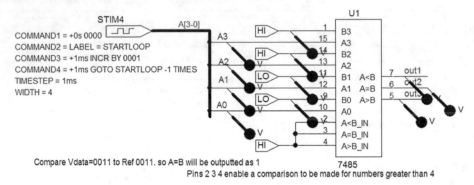

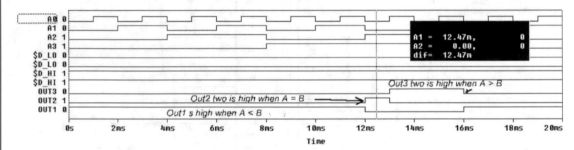

FIGURE 8.10: Digital comparator

FIGURE 8.11: Digital comparator

8.4   DIGITAL COMPARATOR

The circuit in Fig. 8.10 compares two digital numbers. The input digital number is applied with a **STIM1** generator, which outputs four parallel lines to the **A** input lines. This number is then compared to the number setup on the **B** input. We may extend the comparison to higher values by setting pins 2, 3, and 4.

The waveforms for the digital comparator are as shown in Fig. 8.11.

8.5   SEVEN-SEGMENT DISPLAY

A BCD seven-segment decoder/driver display schematic is shown in Fig. 8.12.

PSpice has no LED display part so the next best thing is to rearrange the display resistors as shown in Fig. 8.12. By pressing the current icon after simulating, the seven-segment display action is roughly imitated although you may have to move the current number display a little bit. Placing **HI** and **LO** parts on the input A B C D initiates any input data signal from 0 to 9. A is the least significant digit so that 6 is equivalent to 0110, i.e., A = LO, B = HI, C = HI, and D = 0. The two top resistors RA and RB are simulating an off LED condition, so the other resistors form the number six. (They are arranged in clockwise direction from top RA, RB, RC, RD, RE, RF, and the center RG.) The number of significant digits displayed on the

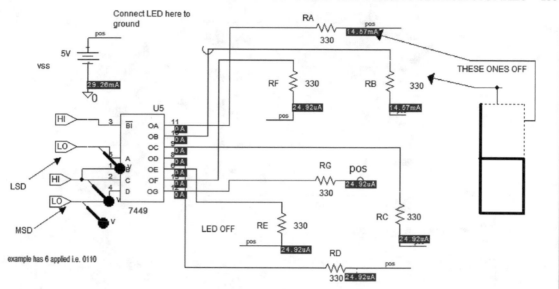

FIGURE 8.12: Seven-segment display

schematic is changed from 2 to 16 from the **Analysis** menu and choosing **Display Results on Schematic**. Unfortunately, we cannot use this technique for dynamic signals but only for high and low parts.

8.6 SIGNAL SOURCES

PSpice has many **DC** and **AC** generator sources. The **AC** sources are classified as analog, digital, and generators that import a text file. The **AC** sources available in PSpice are:

- VEXP,
- VPULSE,
- VPWL,
- VSFFM, and
- VSIN

Digital generators have single and multiple outputs. The **DigClock** generator has a single line output, and the **Stimulus** generators have multiple outputs. The current **IEXP** source generates exponentially growing and decaying current waveforms, with parameters defined:

- i1   Initial current amp none
- i2   Peak current amp none
- td1   Rise (fall) delay sec 0

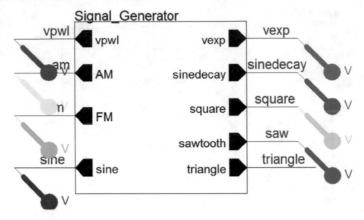

FIGURE 8.13: Function generator

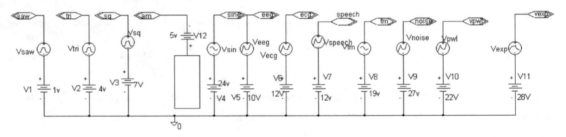

FIGURE 8.14: Inside the function generator

- tc1   Rise (fall) time constant sec **TSTEP**
- td2   Fall (rise) delay sec <td1>+**TSTEP**
- tc2   Fall (rise) time constant sec **TSTEP**

The current i1 for the first td1 seconds decays exponentially from i1 to i2 with time constant tc1. The decay lasts (td2 − td1) seconds. The current decays from i2 back to i1 with a time constant of tc2. The **VEXP** source is similar.

8.7   FUNCTION GENERATOR

The flat modular design function generator in Fig. 8.13 contains a variety of sources. Each source has a series **VDC** part bias placed in series in order to separate out the waveforms in Probe after simulation.

The generator inside is shown in Fig. 8.14.

A display of the different signals, with DC offsets, is shown in Fig. 8.15.

Pressing the **FFT** icon will display the frequency content but you will have to separate the **FFT** displays using **alt PP** key stroke sequences (see index on log command usage).

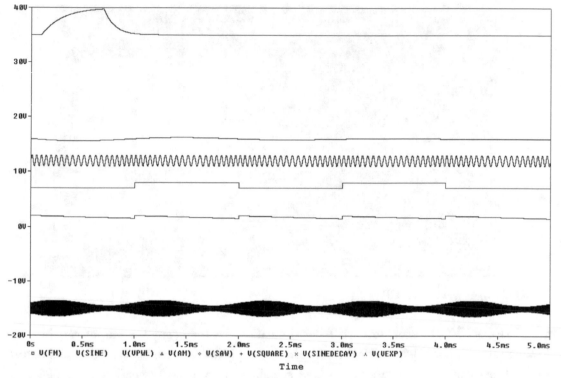

FIGURE 8.15: Function generator waveforms

8.8   IMPORTING SPEECH

The ability to import, export, and process, AUDIO/ECG/EEG files expands the horizon of PSpice greatly. For example, apply a segment of speech to an antialiasing LPF and export the filtered speech to hear how the filter changes the audio signal. Exporting the filtered speech file and playing it back is done using the program '**wav2ascii**' (written by Lee Tobin). The second method uses **Microsoft recorder/Player** for recording and playing back a wavfile and using **Matlab** to produce a text file which is then imported into a schematic using the **VPWL_F_RE_FOREVER** generator.

8.8.1   Wav2Ascii

A very useful program, **Wav2Ascii**, created by Lee Tobin, produces the display in Fig. 8.16. This program creates and writes to disc, a two-second wav file and an ASCII file equivalent called speech.txt containing time and voltage columns. The speech is recorded at 8 bits and sampled at 11025 Hz. Press the **RecordWave/Start** button to record a two-second wav file that is written to the directory where the program is located. Press the **Play** button to hear the wav2ascii.wav speech file. This also creates a text file that may be applied to a schematic using

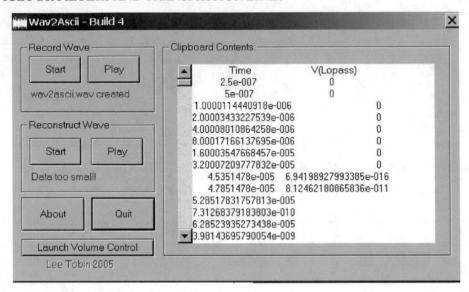

FIGURE 8.16: Wav to ASCII

a **VPWFOREVER** generator part. After simulation, we may copy the output variable from Probe to the clipboard where it is displayed in the Clipboard Contents. Select the **Reconstruct Wave/Start** icon to reconstruct the sound signal. Run the echo simulation (Fig. 8.21), and from Probe, select V(echo) and copy using **Ctrl C**. Press **Reconstruct Wave//Play** to hear the speech and a delay version of it. The **Launch Volume Control** controls the sound level.

8.8.2   Import A Speech File Into A Schematic

The speech file is imported into PSpice using a **VPWL_F_RE_FOREVER** part. The **ABM** filter parts shown in Fig. 8.17 are used to modify the speech. We may use the program Wav2ascii, to listen to the effects of filtering.

From the library, select **VPWL_F_RE_FOREVER,** and set the parameter as shown in Fig. 8.18.

TSF and **VSF** are time and voltage scale factors enabling you to change the voltage and time by any factor you deem necessary. For example, the speech displayed in Fig. 8.19 is quite small so set **VSF** to 5 and simulate again. Set the **Run to time** to 2 s, and **Maximum step size** to 10 μs. Press **F11** to simulate.

Doubling the **TSF** doubles the time scale. Press the FFT to display the speech signal in the frequency domains as shown in Fig. 8.20.

Create a new plot area using **alt PP**. Copy the existing plotted variable **V(RLoad:2)** at the bottom, and paste into the new area using **ctrl V**. This creates two identical plots. From the **Plot** menu, select **Plot/unsynchronize Plot** and then press the **FFT** icon to display the

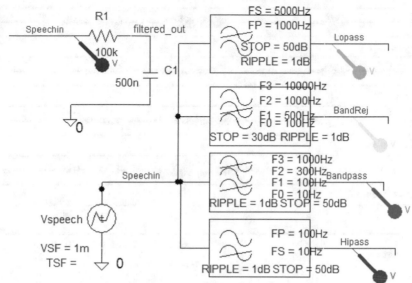

FIGURE 8.17: Importing speech.txt

FILE	C:\Pspice\Circuits\signalsour
TSF	
Value	VPWL_FILE
VSF	1m

FIGURE 8.18: Setting the VPWL parameters

variable in the frequency domain. Investigate the various outputs from the ABM filters and listen to the filter effects on the speech using the Wav2ascii program.

8.8.3 Reverberation/Echo

To illustrate the export facility, draw the schematic in Fig. 8.21 to produce an echo of the original speech (echo is an effect used in the music industry). The delayed speech is mixed with the original signal to enhance the original audio signal. The speech delay is achieved using a transmission line part called **T**. However, this part must be correctly terminated at both ends in a resistance equal to the characteristic impedance Z_0. Select the transmission and set the parameters as shown. Failing to this will result in reflections back toward the source. The delay is specified in the line as {**delay**}. The delay can be then changed in a **PARAM** part.

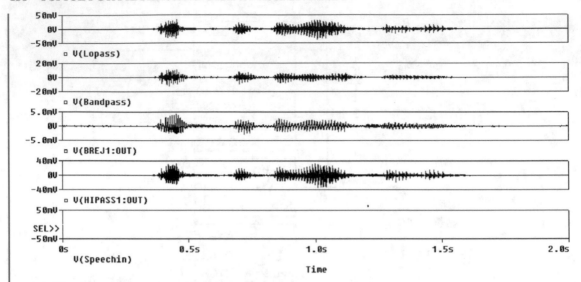

FIGURE 8.19: Filtered speech in the time domain

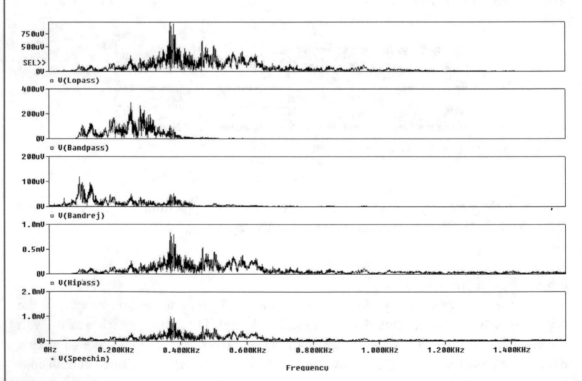

FIGURE 8.20: Filtered speech in the frequency domain

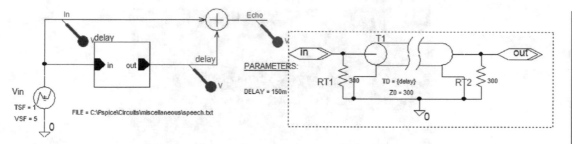

FIGURE 8.21: "Echo unit"

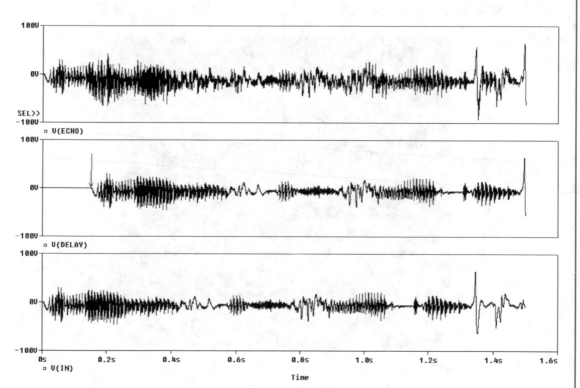

FIGURE 8.22: Time plots of echo speech

Set the **Run to time** to 2 s, and **Maximum step size** to 10 μs. Press **F11** to simulate. Figure 8.22 shows plots of the original and delayed signals. What value of delay does reverberation become an echo (a slight delay between the delay and actual speech)?

Select the variable **V(echo)** and copy using **ctrl C**. Open Notepad© text editor and paste the copied variable **v(echo) (ctrl V)** and save this as a text file to **C:\Pspice\Circuits\signalsources\speech\yourfilename.txt**. From Wav2ascill screen, select **Reconstruct Wave/Start** icon to reconstruct and listen to the processed sound.

FIGURE 8.23: Microsoft sound recorder

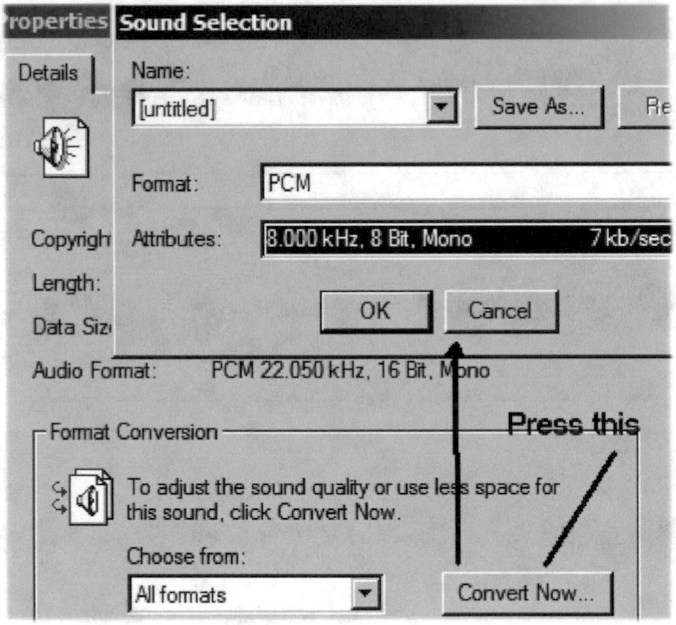

FIGURE 8.24: Properties of sound recorder

8.9   RECORDING A WAV FILE

Create the wav file using the Microsoft recorder shown in Fig. 8.23, and save the file to **C:\Pspice\Circuits\signalsources\speech\speech.wav**. If you don't want to do this, then load the text file "**speech.txt**" onto your hard drive.

Fig. 8.24 shows how the recorder parameters, such as the sample rate etc., may be changed by selecting **File/Properties**. To keep the file small, use a sampling rate of 8000 Hz and 8 bits mono (poor quality).

It is also possible to record a wav file from the Matlab environment. Check the file using a playback recorder such as the Windows Media Player shown in Fig. 8.25.

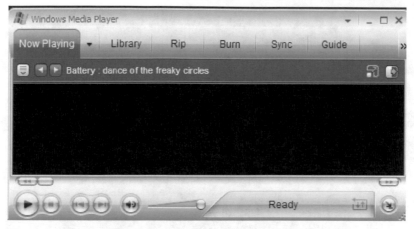

FIGURE 8.25: Windows Media Player

To import speech into PSpice we load the wav file created by the recorder program into Matlab using a sequence of commands from the Matlab double chevron command line ≫ (A better technique is to create a Matlab mfile).

8.9.1   Matlab Code

This Matlab file **right.m** reads in a wav file from C:\Pspice\Circuits\signalsources\ speech\speech.wav

```
**************************right.m*****************************
[soundsamples,fsample] = wavread
('C:\Pspice\Circuits\
signalsources\speech\speech.wav');        % Reads in speech file sampled at
                                          %fsample = 8000 Hz @8bits and returns
                                             the samples and sample rate

soundsc(soundsamples,fsample);            % Listen to the sound
sound = sound(1:8000);                    % one second of speech (64 kbits in total)
time = (1:1:length(sound))/fsample;       % Create a vector of time
time = time(1:8000);                      % one second duration
time = time';                             % Create a column vector
plot(time,sound)                          % Plots in Matlab the speech file
right = [time sound];                     % creates the time-amplitude vector
                                             in two columns

save C:\Pspice\Circuits\signalsources\
speech\right.txt right -ascii;            % save as ASCII (time in first column).
```

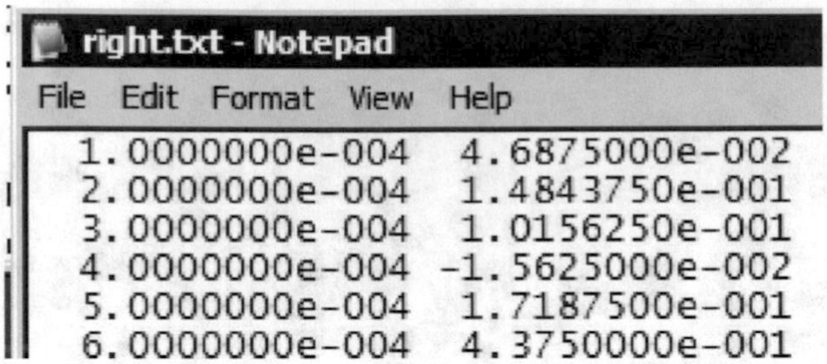

FIGURE 8.26: Speech file in time-amplitude columns

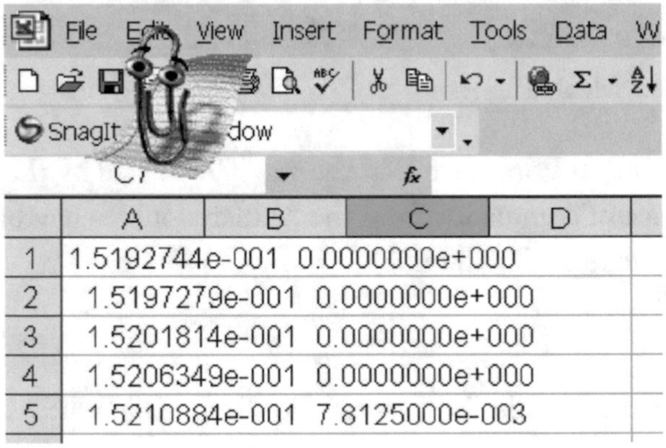

FIGURE 8.27: Importing into Excel©

Fig. 8.26 shows the first column as time and the second column as voltage for the speech file read in from the signal-sources directory using Notepad.

8.9.2 Exporting Speech

PSpice has no mechanism for playing sound files so we must export the speech in order to listen to the effects of any signal processing. For example, we may wish to listen to the effect on the speech file processed by a low-pass filter. To export a signal from the Probe screen, select the trace variable name under the x-axis Probe plot. The selected variable should turn red, so apply **Ctrl C** to copy it and **Ctrl V** to paste to the clipboard. Open up Wordpad© or Notepad© and paste in the copied variable using **Ctrl V**, or **Edit/Paste**. Two columns should appear in the text editor area representing time and voltage (the x and y axis variables), complete with a header on the first line. (Note: Selecting multiple traces and **Copy/Paste** produces the x-axis column and multiple y-axis columns, that may be imported into **Excel** as shown in Fig. 8.27.)

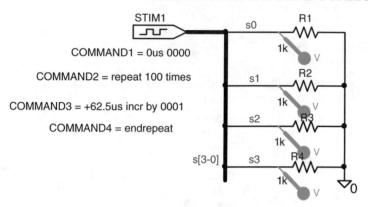

FIGURE 8.28: STIM4 generator

Remove the header row containing the time and voltage names, and save the remaining file content as a **yourname.txt** file. It is now necessary to load this text file into Matlab. Open Matlab and enter the command echoout = load ('C:\Pspice\Circuitse\signalsourcese\speeche\echoout.txt'). When you use **Ctrl C** on the Probe variable, the copied variable contains time and voltage columns. However, we only want the voltage amplitude column, so we have to remove the time column using Matlab.

echoout = echoout (:,2)   %These commands are combined in the mfile echoout.m.

echoout = load('C:\Pspice\Circuits\
signalsources\speech\echoout.txt');   % loads signal from signalsources
echoout = echoout(:,2);   % removes the time column
soundsc(echoout,16000)   % Increase playback sample rate as PSpice performs its own sampling process.

wavwrite(echoout,16000,'C:\Pspice\
Circuits\signalsources\
speech\echoout.wav');   % save as a wav file

When playing the exported signal from PSpice into Matlab, we have to double the sample rate from the original rate.

If you do not want to use Matlab to import and export signals then use the following program.

8.10   EXERCISES

(1) Investigate the use of the STIM4 generator in Fig. 8.28.

(2) Investigate the staircase generator shown in Fig. 8.29.

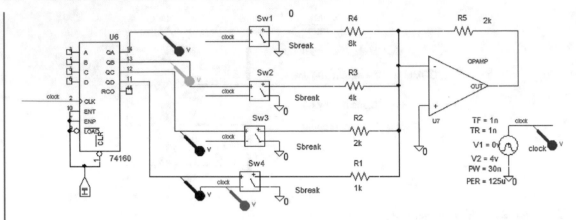

FIGURE 8.29: Staircase generator

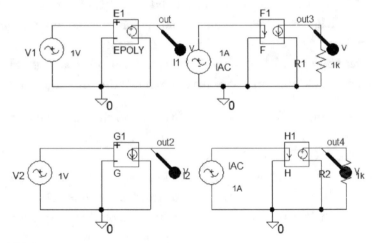

FIGURE 8.30: ABM types

(3) Write and read data to the RAM device. How could this device be used to produce an echo or reverb signal? Hint: Store signals from the ADC in RAM and read them back a short time later and combine with the original signal.

(4) Investigate the **ABM** parts in Fig. 8.30 for use in active filters and two-port circuit simulations.

E = voltage-controlled voltage source VCVS, F = Current-controlled current source ICIS

G = Voltage-controlled current source VCIS, H = Current-controlled voltage source ICVS

TABLE 8.2: Laplace and z-Transform Table

FUNCTION	$f(t)$	LAPLACE TRANSFORM	$f(n)$	$(t = nT = n)$ z-TRANSFORM
Unit step	$u(t)$	$\dfrac{1}{s}$	$u(n)$	$\dfrac{z}{z-1}$
Unit impulse	$\delta(t)$	1	$\delta(n)$	1
Unit ramp	T	$\dfrac{1}{s^2}$	N	$\dfrac{nz}{(z-1)^2}$
Polynomial	t^n	$\dfrac{n!}{s^{n+1}}$	t^n	$\dfrac{T^2 z(z+1)}{(z-1)^2}$ for $n=2$
Decaying exponential	e^{-at}	$\dfrac{1}{(s+a)}$	$e^{-an}u(n)$	$\dfrac{z}{z-e^{-an}}$
Growing exponential	$\dfrac{1}{a(1-e^{-at})}$	$\dfrac{1}{(s+a)(s)}$	$\dfrac{1}{a(1-e^{-an})}$	$\dfrac{z(1-e^{-an})}{a(z-1)(z-e^{-an})}$
Sine	$\sin(\omega t)$	$\dfrac{\omega}{(s^2+\omega^2)}$	$\sin(n\theta)u(n)$	$\dfrac{z\sin n\theta}{z^2 - 2z\sin n\theta + 1}$
Cosine	$\cos(\omega t)$	$\dfrac{s}{(s^2+\omega^2)}$	$\cos(n\theta)u(n)$	$\dfrac{z(z-\cos n\theta)}{z^2 - 2z\cos n\theta + 1}$
Damped sine	$e^{-at}\sin(\omega t)$	$\dfrac{\omega}{[(s+a)^2+\omega^2)]}$	$e^{-an}\sin(n\theta)$	$\dfrac{ze^{-an}\sin(n\theta)}{z^2 - 2ze^{-an}\cos n\theta + e^{-2an}}$
Damped cosine	$e^{-at}\cos(\omega t)$	$\dfrac{(s+a)}{[(s+a)^2+\omega^2)]}$	$e^{-an}\cos(n\theta)$	$\dfrac{z^2 - ze^{-an}\cos(n\theta)}{z^2 - 2ze^{-an}\cos n\theta + e^{-2an}}$
Delay	$f(t-k)$	e^{-sk}	$f(n-k)$	z^{-k}

APPENDIX A: REFERENCES

[1]   Tobin Paul, PSpice for Circuit Theory and Electronic Devices Morgan Claypool publishers Feb 2007.

[2]   Tobin Paul, PSpice for Digital signal processing. Morgan Claypool publishers (Feb 2007)

[3]   Tobin Paul, PSpice for Analog Communications Engineering Morgan Claypool publishers Feb 2007.

[4] Theodore Deliyannis, Yichuang Sun, John Kelvin Fidler, Continuous-time Active Filter Design CRC press 2000.

[5] Don Lancaster, Active Filter Cookbook, Synergentics press USA.

[6] Tobin Paul, PSpice for Digital Communications Morgan Claypool publishers Feb 2007.

Index

Author Biography

Paul Tobin graduated from Kevin Street College of Technology (now the Dublin Institute of Technology) with honours in electronic engineering and went to work for the Irish National Telecommunications company. Here, he was involved in redesigning the analogue junction network replacing cables with PCM systems over optical fibres. He gave a paper on the design of this new digital junction network to the Institute of Engineers of Ireland in 1982 and was awarded a Smith testimonial for one of the best papers that year. Having taught part-time courses in telecommunications systems in Kevin Street, he was invited to apply for a full-time lecture post. He accepted and started lecturing full time in 1983. Over the last twenty years he has given courses in telecommunications, digital signal processing and circuit theory.

He graduated with honours in 1998 having completed a taught MSc in various DSP topics and a project using the Wavelet Transform and neural networks to classify EEG (brain waves) associated with different mental tasks. He has been a 'guest professor' in the Institut Universitaire de Technologie (IUT), Bethune, France for the past four years giving courses in PSpice simulation topics. He wrote an unpublished book on PSpice but was persuaded by Joel Claypool (of Morgan and Claypool Publishers) at an engineering conference in Puerto Rico (July 2006), to break it into five PSpice books. One of the books introduces a novel way of teaching DSP using PSpice. There are over 500 worked examples in the five books covering a range of topics with sufficient theory and simulation results from basic circuit theory right up to advanced communication principles. Most of these worked example circuit have been thoroughly 'student tested' by Irish and International students and should mean little or no errors but alas... He married Marie and has four sons and his hobbies include playing modern jazz on double bass and piano but grew up playing G-banjo and guitar. His other hobby is flying and obtained a private pilots license (PPL) in the early 80's.